SCIENCE EDUCATION FOR AUSTRALIAN STUDENTS

SCIENCE EDUCATION FOR AUSTRALIAN STUDENTS

TEACHING SCIENCE FROM FOUNDATION TO YEAR 12

Edited by **ANGELA FITZGERALD**
and **DEBORAH CORRIGAN**

Routledge
Taylor & Francis Group
LONDON AND NEW YORK

First published 2018 by Allen & Unwin

Published 2020 by Routledge
2 Park Square, Milton Park, Abingdon, Oxon OX14 4RN
605 Third Avenue, New York, NY 10017

Routledge is an imprint of the Taylor & Francis Group, an informa business

NATIONAL LIBRARY OF AUSTRALIA

A catalogue record for this book is available from the National Library of Australia

Internal design by Romina Panetta
Set in 12/17 pt Adobe CaslonPro by Midland Typesetters, Australia

ISBN-13: 9781760296889 (pbk)

Contents

Figures and tables *vii*
Contributors *ix*
Introduction xiii

Part I Science and the learner

1 Becoming a teacher of science: Introducing a journey of identity and evidence 3
Angela Fitzgerald and Deborah Corrigan
2 What the learner brings 26
Kathy Smith and Angela Fitzgerald
3 The role of context and relevance 57
Deborah Corrigan and Angela Fitzgerald

Part II Understanding science

4 The science curriculum 89
Deborah Corrigan and Karen Marangio
5 The nature of science 113
Kathy Smith, Deborah Corrigan and Jennifer Mansfield
6 Science investigations 143
Deborah Corrigan, Angela Fitzgerald and Gillian Kidman

Part III Pedagogy and assessment

7 Assessment for student and teacher learning 179
Debra Panizzon and Stephen Keast

8 Pedagogical content knowledge 205
John Loughran

Part IV New approaches in teaching and teacher development

9 Transitioning into professional practice 235
Kathy Smith and Rebecca Cooper

10 Science education in informal settings 262
Alan Reid and Ben Liu

11 Science and STEM education 302
Deborah Corrigan and Greg Lancaster

Index *321*

Figures and tables

FIGURES

1.1	Fortune line of three students' feelings about studying forces	11
3.1	Actions necessary for ways of knowing science in contexts	58
3.2	The 'ways of knowing' framework	59
4.1	An advanced organiser for the intended, implemented and experienced curriculum	93
4.2	The three-dimensional Australian Curriculum	105
6.1	Flowers and foliage of *A. glabrata*	145
6.2	Positions of variables in a table	161
7.1	Simple concept map	187
7.2	Concept map of increased complexity	188
7.3	Venn diagram	189
10.1	Education for Conservation: Seal the Loop Action Day—schools undertaking community conservation campaigns in their local community	264
10.2	Education Resource Centre Library with giraffe pelt	271
10.3	Zoo School: Inside the first Zoo School classroom at Melbourne Zoo	272
10.4	Zoo School: The first Zoo School classroom at Melbourne Zoo, now The World of Frogs	273
10.5	Teaching in the Zoo School: Zoo educator with students and animal encounter	274
10.6	Student feeds a tawny frogmouth, Healesville Sanctuary	274

10.7 Education for Conservation: Conservation focused animal encounters—students meet a mountain pygmy possum 275
10.8 Zoo School—towards less formal interactions? 276
10.9 Secondary students inspired by a lion keeper talk, Werribee Open Range Zoo 276
10.10 Secondary students planning the locations of their inquiries, Werribee Open Range Zoo 282
10.11 A zoo-based conservation organisation: Zoos Victoria Fighting Extinction 284
10.12 Education for Conservation: Out of the classroom exploring the zoo in small groups 286
11.1 Rule-based problem-solving 309
11.2 Converting a rule-based problem to a non-rule based problem 310

TABLES

3.1 Stakeholders and relevance 79
4.1 Focus and unifying ideas at different stages of schooling 107
5.1 Problems that needed solving before work could begin 116
5.2 Some values of science and their meaning 126
6.1 Characteristics of investigation types 153
6.2 Examples of differing investigation questions 154
6.3 Substances provided for density experiment 169
7.1 Main types and purposes of assessment 184
8.1 An abbreviated CoRe illustrating the big ideas (horizontal axis) and pedagogical prompts (vertical axis) 218
9.1 Stages of career and their characteristics 246
9.2 AITSL Professional Standards 259
9.3 ASTA Professional Standards 251
10.1 Types of education program according to participant choice and goals and provider goals 278
11.1 STEM as a way of thinking and acting 312
11.2 Summary of approaches to curriculum integration 315

Contributors

Rebecca Cooper teaches general science and physics specialism units. Her research is focused on the career progression of science teacher educators and the development of science teachers' pedagogical content knowledge and pedagogical knowledge.

Deborah Corrigan was a chemistry and biology teacher before becoming a science education researcher in teacher education. Her research focuses on science that is relevant and useful to learners and she explores the values that underpin knowledge in science and STEM more broadly.

Angela Fitzgerald has been involved in science education in primary and secondary settings as a teacher and teacher educator for fifteen years. Her research mainly focuses on supporting primary pre- and in-service teachers in developing their confidence and competence in science learning and teaching.

Stephen Keast is a retired science academic currently teaching physics to senior secondary students. His research interests were in teacher learning, pedagogical reasoning and teacher professional wisdom. In returning to the classroom, he is implementing much of his research in the act of teaching.

Gillian Kidman has research and teaching interests in science and mathematics. She is particularly interested in inquiry-based processes, and the potential inquiry has for the integration of science with other disciplines such as the development of STEM. She is currently working on the transdisciplinary nature of STEM inquiry.

Greg Lancaster has extensive experience in teaching science and physics in the secondary and tertiary sectors in Victoria. His research interests include the development of conceptual understanding in physics, the communication of science and science teacher professional learning.

Ben Liu is an informal science educator who works as the learning engagement manager within the Wildlife Conservation and Science team at Zoos Victoria. His teaching and learning experience focuses on the development of conservation education experiences for a range of audiences.

John Loughran was a high school science and geography teacher for twelve years before accepting an appointment as a teacher educator in the university sector. His research interest revolves around the development of teachers' professional knowledge, in particular the field of science teachers' pedagogical content knowledge.

Jennifer Mansfield is a lecturer in primary science and secondary biology method. Her passion for science education has seen her work as a scientist and secondary school science teacher. Her research focuses on teacher professional learning and development, including her own, which she researches through self-study.

Karen Marangio's work revolves around psychology and science teacher education. Her research interests include teachers' views, values and teaching of psychology and science in schools, science curriculum, co-teaching and developing the pedagogy of pre-service teachers.

Debra Panizzon is an adjunct at Monash University and research analyst for the Teachers Registration Board of South Australia. Her research interests include STEM policy, assessment in science and mathematics education, cognition and learning and rural education, with much of this work involving teacher partnerships.

Alan Reid works in the area of environmental and sustainability education. His research focuses on teachers' thinking and practice in relation to traditions, capacities and issues in environmental and sustainability education theory and research. He edits the international journal *Environmental Education Research*.

Kathy Smith specialises in primary science education and teacher professional learning. She began her career as a primary classroom teacher and has been involved in science education for the last 30 years. Her research interests include understanding the conditions that promote quality teacher professional learning.

Introduction

Becoming a teacher is complex and challenging work—a reality that doesn't disappear with the completion of your teacher education study. Grappling with who you are as a teacher and what characterises your practice will most likely be an ongoing process throughout your career. You will perhaps find, though, that your focus shifts over time and with experience from concerns connected with lesson planning and classroom management—common considerations for early career teachers—to thinking more broadly about the range of components that contribute to quality learning and teaching. While it might be difficult to imagine this in your current position as a pre-service teacher, the intention of teacher education is to prepare you for your career as an educator, not just your first few years in the classroom. With this longer-term vision in mind, this book aims to push your thinking about what matters in science education by planting the seeds of ideas from which you can work to open up possibilities now and into the future.

In encouraging you to think differently about science learning and teaching, the contributors to this text—the Science Education Research Group (SERG) in the Faculty of Education at Monash University—are drawing on a long and rich history in this field. SERG has research experience in science education spanning over 50 years, resulting in a significant evidence base upon which to draw. The insights and learnings that came from this background sparked the development of this book. As the team was becoming increasingly aware of the need to better support teachers at all stages of their careers to draw upon current research to inform their science learning and teaching practices, this format seemed like an ideal and meaningful way to achieve such a goal—particularly in terms of reaching pre-service teachers.

This collection does not pretend to be prescriptive or definitive. It has been designed as a research-based book that acts as a companion to the science education you will experience as part of your teacher education program. Uniquely, it tackles key pedagogical ideas in science education that have relevance across the learning continuum from the beginning of primary school through to the senior years of secondary schooling. Throughout this book, you will encounter numerous examples and cases drawn from decades of research and experience in the field that provide practical insights into science learning and teaching from Foundation (Prep) to Year 12. At times, you might wonder about the relevance of these stories in relation to your future practice, but we encourage you to consider these insights for what they are and think about the key learnings you can take away to apply in your own context.

Thinking differently about what science education is and might look like is not easy. In choosing to push your thinking (and essentially your future practice), this book may challenge your existing beliefs as well as your lived experiences of science during your own schooling. It may also raise more questions than it provides answers about science education to start with, and it could well require you to move outside your comfort zone. However, ultimately, opening up your mind to the possibilities that exist will enable you to develop into a teacher who, informed by evidence and current research, can enact quality and contemporary science learning and teaching practices.

Angela Fitzgerald and Deborah Corrigan
February 2018

PART I

SCIENCE AND THE LEARNER

The decision to start this book with a focus on the learner and learning is a deliberate one. While as a pre-service teacher you may be understandably anxious to focus on the craft of teaching science, it is important to recognise one of the key purposes driving our work as teachers of science—to engage and support students in learning science. Equally, we need to remember that we are also learners of science and to value this learning process. By putting learners and learning first, we are making a statement about what ultimately matters in science education and encourage you to keep this in the forefront of your mind as you continue in your learning to teach journey.

Part 1 teases out two key areas for consideration:

1 *learner backgrounds*—understanding that the range of existing ideas, beliefs and perspectives held by learners—both students and teachers—impacts directly on the ways meaning in science is constructed
2 *context and relevance*—identifying how and why these two constructs have a critical role to play in supporting and enabling authentic and meaningful science learning to take place

The part begins by getting you to consider yourself as a teacher of science and to think about your own identity, as well as providing a guide to the structure of the book as a whole.

This section of the book is valuable because it will push you to think more deeply about the learning of science (for students and for yourself) and the role that you, as a future teacher of science, will play.

CHAPTER 1

BECOMING A TEACHER OF SCIENCE: INTRODUCING A JOURNEY OF IDENTITY AND EVIDENCE

ANGELA FITZGERALD *and* **DEBORAH CORRIGAN**

LEARNING OUTCOMES

After reading this chapter, you should be able to:

- consider what quality learning and teaching in science might look like over the course of your career, including as a pre-service teacher
- start to consider the notion of teacher identity and how you connect with this as a future teacher of science
- draw on relevant research to inform your practice as a teacher of science
- understand the scope of this text and the focus of each chapter as an evidence-based collection.

LOOKING BACK TO LOOK FORWARD: A REFLECTION

It doesn't take long for me to scan my memory and remember what it was like being a pre-service teacher. It might have been fifteen years ago since I was in my final year of a Bachelor of Education (Secondary) course, but I remember it—maybe not like yesterday, but certainly last week. When I do start looking back, a lot of things start flooding in. On reflection, it strikes me that it was a jam-packed year and a steep and at times intense learning curve. Outside of my ten weeks of placement that year, there is one particular memory that looms large over the others. It was a peer-teaching task for our general science method unit. The idea was to work with another pre-service teacher to develop and implement a lesson using a learning outcome chosen from the CSF II (the Curriculum and Standards Framework II, which was the Victorian Curriculum of its time). It was one of our assessable tasks for that semester for a pass grade only. I don't remember my teaching partner's name, but I do clearly remember that we totally threw ourselves into the task with the aim of creating an engaging and informative lesson. We decided that our lesson would be the opening in a learning sequence on the respiratory system. Preparation seemed to take place over weeks, trawling so many resources both about the respiratory system and ways to teach it. With all the information we needed to inform our planning, we launched into putting all the pieces together to form a coherent learning journey through the respiratory system. Our students/peers were to participate in a role-play in which they would take on the role of one of the particles that make up the air we breathe. Wearing a badge identifying who they were (around 70 per cent of the group were nitrogen, 20 per cent were oxygen and so on), they passed through the 'nose' (the door to the classroom) to enter into the respiratory system—but not before becoming entangled in streamers (nose hairs), being blasted with heat from a hair dryer (warming environment of the nasal cavity) and sprayed with water from a vaporiser (mucus). Once through the nose, our students continued their active participation in the role-play as they passed through a number of parts of the body, including the trachea, bronchi and lungs. Alongside the lived experience of the role-play, there were a couple of activities that further demonstrated how the respiratory system works. I have no idea how our lesson ended, but I remember being energised and excited by how the session had gone as we quickly packed all our resources into a box to prepare for the next peer-teaching session—but this time as learners.

—Angela Fitzgerald

FROM PRE-SERVICE TO IN-SERVICE TEACHER OF SCIENCE

Perhaps you see something of yourself in this reflection. In your time as a pre-service teacher, you too will spend hours planning for one lesson with the focus on creating engaging science activities in which learners can participate. Our experiences as teacher educators, across primary and secondary science, tell us this is more than just a common story but an expectation. Research also documents that this activity focus is a very normal experience (Appleton, 2002). In learning to be a teacher, it is understandable that your focus might be on your role as a teacher and the actual act of teaching rather than on the learner and their learning. After all, your experiences of education up to this point, as both a student and a pre-service teacher, have essentially been on observing and thinking about what the teacher is doing.

However, your own experiences of science as a learner are significant in shaping your approach to science education as a teacher. You might be passionate and knowledgeable about a particular area of science, or perhaps you found learning science to be uninteresting and disconnected from your life experiences. Regardless of which end of this spectrum you connect with, there is a natural inclination for pre-service teachers to want to be the kind of science teacher who is seen by their students as fun and dynamic. Delivering a range of hands-on activities connected with a conceptual area in science can be perceived as achieving this goal. While the opening reflection certainly brings these things into focus, in using a more critical lens to interpret this experience, it might be pertinent to ask whether this approach results in quality science education.

To start to unpick an answer to this question, we need to shift gears for a moment and consider what evidence tells us. The first author of this chapter (Ange) completed a PhD project (Fitzgerald, 2010) exploring what quality science learning and teaching looks like in a classroom setting. While this research was positioned within primary school science education, it certainly uncovered findings that have applicability to both primary and secondary classrooms. In making sense of the existing body of knowledge, research into quality science learning and teaching can broadly be identified as focusing on what influences teaching practice and how this practice then impacts on student learning in science. Perhaps the most important finding from this work is that where and how learning and teaching are positioned are highly contextual. While this might seem completely obvious, it is surprising the number of times we—as pre-service and in-service teachers—forge on in our practice without taking the time to consider the context in which we find ourselves, including particular contextual factors, our beliefs and our knowledge (Fitzgerald, 2012). Together, these three constructs have a considerable impact on what we might consider as contributing to quality science education.

Contextual factors such as the socioeconomic status of a school community, our students' ages and their cultural backgrounds will influence the nature of quality science education practice. For example, a school where a large number of students are new immigrants may have a different set of expectations for student learning than one where most of the students are from the homes of third-generation Australians with professional parents who own their own homes. It is acknowledged that teachers hold beliefs about science and science education that have evolved during their lifetimes (Pajares, 1992). For example, many current teachers learnt and still believe that Pluto is a planet. If they also believe that science is about proven facts, then they will have some difficulty reconciling Pluto's downgrade from planet status, which occurred in 2006 when the International Astronomical Union determined that Pluto was now a 'dwarf planet'. The nature of these beliefs also has an influence on the practices of quality science teachers (Duschl, 1983). For example, if teachers believe that the **scientific method** represents the way science is conducted, they will have great difficulty embracing science investigations that are descriptive, **correlational** based, in fieldwork or in multidisciplinary teams—which is how much of the work of science is currently conducted (see Chapter 6 for more detail). Similarly, teachers' knowledge about scientific concepts, skills and the nature of science, alongside their knowledge of science teaching, has a direct influence on classroom actions and behaviours (Harlen, 1997). For example, do you as a teacher accept that all your students will bring some scientific knowledge to your classroom and shape your teaching approaches to encompass their ideas? The knowledge base, belief systems and contextual factors that surround us influence our ability to enact quality science learning and teaching, and shape why we might teach and learn in the ways we do.

scientific method described in simple terms as formulating a rigorous hypothesis and then testing it. For more discussion on the concept see Chapter 6

correlational a science investigation that tries to establish the relationship between two things

We cannot, and should not, ignore a number of other factors that make a significant contribution to quality science education. For example, learning and teaching processes and activities act as the bridge between teacher practice and the engagement of students in the learning of science. Like teachers, students also bring ideas and understandings that they have developed through their experiences of the natural world and throughout their schooling to the learning of science. Particular educational acts such as opportunities to engage in talk (Mortimer & Scott, 2003) and to use different representational forms to make sense of science concepts can also act to strengthen student understandings (Prain & Waldrip, 2006). Ultimately, though, for these opportunities to make a difference, students' science learning needs to be monitored and assessed with regular feedback provided by the teacher (Cowie, 2002). All these factors and considerations might be bundled into a neat package here, but it is important to acknowledge that learning and teaching are complex and inherently messy. And perhaps when you are striving to achieve quality, in science education it starts to become even more so.

As a pre-service teacher, it is possible that you may be starting to feel overwhelmed by these insights. Transitioning from where you find yourself now to being an in-service teacher implementing quality science learning experiences may seem like a long road to tread. It is easy to start feeling impatient to get to this point as a teacher—especially when you are coming to the end of your initial teacher education course. There is a real sense of wanting to get out there into your own classroom, away from the theoretical and abstract aspects of university education, and put all the things you learnt about teaching into practice. While enthusiasm is a great starting point and experience is a great teacher itself, these two things will only get you so far. The transition from being a pre-service to in-service teacher is relatively straightforward—you become employed as teacher. It is the shift to quality that requires dedication and a commitment to ongoing learning and growth.

The good news is that it is possible, as a pre-service teacher, to start to prepare yourself for this longer haul. First, you can start work on developing your sense of identity as a teacher, and more specifically as a teacher of science. Second, you can start to think more deeply about what you do as a teacher of science and why, and support such decisions by using evidence to inform your practice. While the intention of this chapter is to introduce and orientate you to this book, it is also to unpack these two points in more detail and highlight why they matter in terms of your future practice. This chapter also intends to convey why this edited collection of chapters matters in terms of preparing you as a future teacher of science by orientating you to the contemporary issues and practices underpinning quality science learning and teaching in Australian primary and secondary classrooms.

WHO ARE YOU AS A FUTURE TEACHER OF SCIENCE?

> **Reflections from a pre-service teacher**
>
> I don't really know who I am as a teacher yet. I think I know, well I know what kinds of things I'd like to do in my classes to make learning fun for my students. I guess I just haven't really had a chance to put all the things that I have learnt about at uni and all my thinking about teaching into action yet in the ways that I would like. I have had some great placement experiences and learnt so much about teaching from those times, but in reality it is more about fitting into your mentor teacher's classes, using their lesson plans and kind of sticking to their teaching style and approach more than your own. (Nick, pre-service secondary teacher)

The notion of who you are as a person is quite possibly something with which you have grappled, and will continue to grapple, over your lifetime. But regardless of where you are in answering this existential question for yourself, if you were asked to articulate who you are, you would no doubt be able to pick up on some key features that have shaped and influenced you to become the person you are today. The above quote from Nick captures something different, though. It starts to pick up on the multiple identities that we might carry around with us and occupy at different times (Kamler & Thomson, 2006). As future teachers, you have a new identity taking shape that begins during your initial teacher education years—who you are as a teacher. This book pushes this further by encouraging you to consider your identity through the lens of being a future teacher of science.

It is important to recognise that your identity as a teacher will not be static; it will not stay the same over your career. Your identity and how you see yourself as a teacher will be dynamic and changing, moving with you as you reflect on and learn from the different experiences you have. It is worth pausing for a moment to consider how you might articulate who you are and how you view yourself as a pre-service teacher. Possibly words like 'learner', 'learning', 'student-focused' and 'passionate' might come to mind. In revisiting the reflection that opens this chapter, these terms certainly might have mirrored how the author viewed themself as a pre-service teacher. Now consider this notion again, but in terms of how you think of yourself as a pre-service teacher of science. Does anything change? For some readers, it might not; for others, it will. It is interesting to consider this change and what it might look like. Additions to your thinking might include terms like 'knowledgeable', 'confident', 'engaging' and 'experienced', although it might equally draw reflections from the other end of the continuum. It is important to recognise that there are no right or wrong answers. Your perceptions of yourself and your identity hold true to you. It is more important to remember that these perceptions can change and evolve. Even as you engage with the different chapters in this book, you may begin to think and position yourself differently in terms of who you are as a pre-service teacher of science.

Just as being a pre-service teacher of science brings particular ways of thinking about science education and your place in it, so does being a student or teacher of science. During your time as a student, you will have engaged with and learnt about science in various ways and at different points. This may have made up part of your compulsory study (up to Year 10), but you may also have opted to pursue science education in the post-compulsory years. Just as they do when you are a pre-service teacher of science, your perceptions of yourself as a learner of science most likely also had a sense of ebb and flow. In drawing on the notion of a continuum again, you may at times have felt connected, engaged and inspired through learning science and at other times disconnected, unengaged and uninspired—or any combination in between. This will have been due to a range of possible factors, such as the learning approach, the area of science and level of achievement.

For example, in Figure 1.1, a fortune line maps the feelings of three students, John, Sue and Ashir, studying forces (the push and pull on objects coming from the objects interacting with each

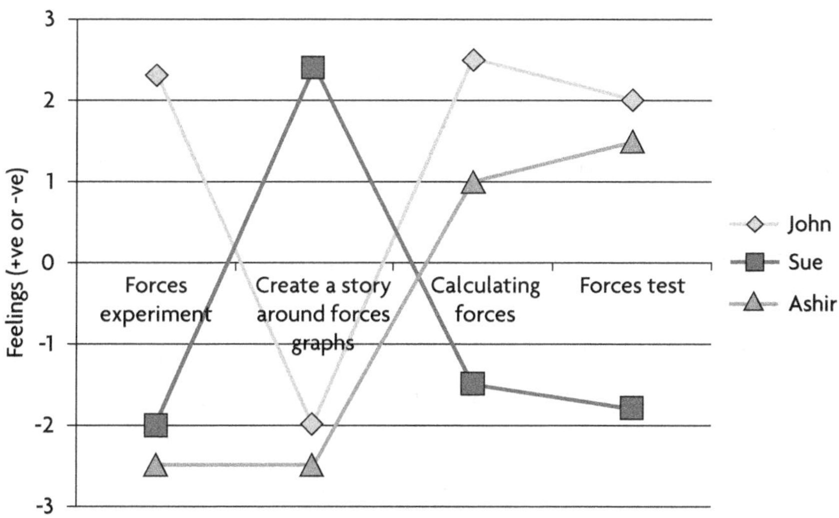

Figure 1.1 Fortune line of three students' feelings about studying forces

other). Their feelings with respect to events through the course of their study are tracked as being either positive (indicated by a positive value between 0 and 3) or negative (indicated by a negative value between –3 and 0). These three students have responded very differently to the learning events of a forces experiment, creating a story around some graphs of forces, calculating the size of forces and a forces test.

Regardless of how these experiences are categorised or ranked, they have a significant impact on shaping how you perceive and situate yourself as a learner of science, which in turn can influence the ways we position ourselves as teachers of science. While, as learners, we may have been able to state that we 'don't do science' or don't like particular aspects of this learning area, as a teacher you won't have this luxury. Likewise, you may have a passion and an in-depth knowledge of science (or a certain branch), but your challenge will lie in supporting students who don't feel as connected to science to find its relevance to their lives. Essentially, regardless of how you might perceive yourself as a learner of science, being a teacher of science requires a different persona—one that isn't overshadowed by personal (dis)interest or a (dis)connection with science.

Being a teacher of science is a not a cookie cutter role—it is not going to look the same for everyone or at every stage of their career. Again, there needs to be an awareness about the dynamic and changing nature of what being a teacher of science might look like and what it might mean to you. While there certainly are some commonalities across science education and the approaches used to promote quality science learning, there are numerous differences that are driven mainly by learners and their learning needs. Broadly, science education has a responsibility to develop the scientific literacy of all students (Goodrum, Hackling & Rennie, 2001), so they are able to operate in the world as science-savvy citizens. A variety of approaches will be needed to attain a goal like this. For example, the ways in which primary school students are prepared to be scientifically literate are very different from those acceptable to secondary school students. Regardless of these differences, an overarching tenet is the importance of growing competence in dealing with greater complexity over time. This is not just about conceptual understandings, but the types of skills and attributes that are drawn upon to inform these understandings—for example, observations, using evidence to support a claim, considering the values that underpin decision-making and so on.

As an early years teacher of science, you would be supporting young children to draw upon their observations of the world to help them make sense of the world and inform their decision-making. For example, young children can articulate their understandings of what is happening with the weather (sunny, rainy, cloudy) and use this knowledge to assist them in making decisions about what to wear (it is sunny, so I will need a hat but not a jacket). These are steps towards using appropriate language to describe the science phenomenon that is taking place and the impact it has on our lived experiences. As a primary school teacher of science, you would be supporting the children with whom you work to draw further on evidence from a range of sources—not just observations—to support their claims about the experience they encounter in their daily lives. For example, primary children could draw on their experiences with shadows and modelling to explain that the sun doesn't go to 'sleep' at night, but that the Earth is rotating and that it is night because a particular part is no longer facing the sun. At this level of schooling, science becomes even more of a lens for making sense of what is happening in the world and why. As a secondary school teacher of science, you would be supporting students to grapple with increasingly more abstract science concepts and applying the relevant patterns, rules and theories to bring meaning to how and why things happen in the way they do, and to assist in predicting whether or not something may occur. For example, adolescents can apply the laws of physics—particularly in relation to force and momentum—to gain informed insights into why it is necessary to maintain a safe braking distance from the car in front. At this stage, science increasingly gets drawn upon as a way of knowing, thinking and acting as a student of science as well as a citizen.

You may have noticed that throughout this chapter we have referred to you being a 'teacher of science' rather than a 'science teacher'. It may seem just a matter of semantics, but there is a lot of meaning in this subtle variation. The important part of this phrasing is noting what we are positioning up front. By using 'teacher of science', we are placing a value and emphasis on the teaching aspect of your future work. While you may enter the classroom with expertise in a particular area or field of science, equally you may not. You will, however, be starting your career as a teacher with expertise in teaching, and this is the aspect that matters most in this context. At different times you may teach different content areas, both within and outside the sciences, but it will be the quality of the learning and teaching occurring that will be the primary focus. As a teacher, you will be responsible for drawing on a range of pedagogies and approaches to best meet the needs and interests of the learners you are supporting. Through doing this, you will ultimately be enriching and enhancing their learning. This is not to infer that engaging with science and building your own scientific understandings and knowledge, along with your students, is not important, but it is your role and expertise as a teacher rather than a scientist that will be brought to the fore through your practice and burgeoning identity.

This section, as will be reflected through the rest of this book, has started to provide you with ideas about what it might mean to identify as a teacher of science. The intention was also to challenge your notions of yourself as a future teacher of science and to start to consider how this positions who you are and what you do to foster quality learning and teaching in the classroom. The next section starts to consider how evidence might inform this by examining what each chapter will contribute to this developing narrative.

USING EVIDENCE TO SUPPORT YOUR TEACHER OF SCIENCE JOURNEY

> **Reflections from a graduate teacher**
> I want to continue learning about teaching and I would like to use research to do that, but I have no idea where to start. I mean . . . it is hard to know what ideas I should be paying attention to and what research really matters for new teachers. And what is quality? I have no idea about how to really determine that either! There is a lot of stuff out there. Even harder is working my way through a journal article and then finding that I have no idea about how to apply its findings to my class or my own teaching. Why can't it be more straightforward? I want to be a teacher who is informed by research, but education researchers are not making it easy for me.
> (Leila, graduate primary teacher)

This quote from a graduate teacher laments the struggles she faces in her desire to continue to learn about teaching in a way that is informed by research, but applicable to her practice. There seems to be a middle ground missing in the literature. At one end of the spectrum, there are numerous how-to texts packed with practical ideas and activities; at the other end, there is academically rigorous and focused work that seems removed from the day-to-day experiences of the classroom, with very little in between. This book is an attempt to redress this disequilibrium. Each chapter engages with research and literature—both established and original—drawn from different areas of science education, and pushes the boundaries a bit further to present a contemporary approach to science learning and teaching across the primary and secondary years. You will find margin definitions peppered throughout the book that will provide you with further insights about key concepts and constructs in both the areas of science and science education. It is important to note that the evidence base underpinning this book is academically sound, with the innovation lying in the interpretation and presentation of research to make sense both practically and theoretically for pre-service teachers—future teachers of science.

Each chapter draws on relevant research—again, both established and original—to inform the messages it is sharing with you, and to bring this work to life through illustrative quotes, case studies, reflections and examples. There is a particular focus on translating this evidence base into a usable form by teasing out the implications for practice as well as highlighting possible strategies or approaches as appropriate. The remainder of this section provides an overview of each chapter, highlighting the theoretical and/or conceptual ground covered alongside the boundaries being pushed to bring relevance to this work for pre-service teachers.

constructivism a theory of learning based on the ideas that knowledge is constructed by the learner and that learners are active in the process of meaning-making

alternative conceptions the conceptions of science held by students that are not scientifically accurate. Often incorrectly identified by the term 'misconceptions'

Chapter 2: What the learner brings (Smith and Fitzgerald) launches into this text by bringing to the fore three key constructs that significantly inform quality science learning and teaching: the theoretical lens of **constructivism** (Claxton, 1990; White & Gunstone, 1992); the notion of **alternative conceptions** (Wenning, 2008) and the impact they have on science understanding; and different ways of knowing science through culturally constructed meanings. Smith and Fitzgerald were prompted to tease out these constructs after they received an email from one of their pre-service teachers, which raised questions about how these ideas had been represented in one of their primary science education classes. Using the ensuing series of emails as well as a range of vignettes from primary and secondary settings, this chapter pushes the boundaries of the existing body of knowledge to ensure that it is meaningful to future teachers of science and provides insights into the implications for practice.

Chapter 3: The role of context and relevance (Corrigan and Fitzgerald) picks up the key themes of authenticity and meaningfulness that emerged from the previous chapter by exploring the role of context and relevance in science learning and teaching. Informing this work is the significant research base from which Corrigan, Buntting and Jones (2013) have drawn to develop a framework representing science knowledge along two continua ranging from individual concepts to big ideas along one, and simple applications to complex contexts along the other. The intention of this frame is to support the planning and implementation of science learning experiences that are contextually situated and relevant to the learner. With this framework in mind, two case studies written by teachers—Gillian (primary school) and Diane (secondary school)—became the focus of this work. Analysis of their insights into their science teaching practices resulted in the development of different themes that identified a number of considerations and conditions contributing to how relevant (or not) the science learning was to the context and the learners.

Chapter 4: The science curriculum (Corrigan and Marangio) starts to position the constructs that have been unpacked in the earlier chapters within Australian curriculum frameworks and policies. The innovation in this work is evident in the creation of an advanced organiser, which teases out the notions of curriculum being intended, implemented, and realised or experienced in practice in ways that make sense to teachers of science. Using a set of questions as a guide and a teacher-written case study to further illustrate the critical role of the teacher in navigating curriculum, this chapter unpacks why curriculum looks the way it does and what this means for schools, teachers and learners. This showcases the purpose and possibilities rather than just the pitfalls (Reid & Price, 2018) to encourage teachers of science to use curriculum as a guide to create meaningful learning experiences in science.

Chapter 5: The nature of science (Smith, Corrigan and Mansfield) raises further questions about relevance and context by considering the gap that exists between how scientists practise science and how science is represented in schools. Drawing on a case study that documents the trials and tribulations of a research scientist, Professor Beth McGraw, and her team as they conduct research into dengue fever, this work starts to tease apart what the nature of science (NoS) actually is (Lederman, 1992, 1996), and what it means for science learning and teaching. This is done in a way that highlights the implications for practice while positioning this construct as a vehicle not just for developing future scientists, but for the creation of more science-savvy citizens. This chapter challenges the reader to move beyond a reliance on dispensing science content knowledge to integrate content in ways that enable learners to critically evaluate scientific evidence and apply it in meaningful contexts.

Chapter 6: Science investigations (Corrigan, Fitzgerald and Kidman) is a further acknowledgement that scientists are not driven solely by content knowledge, but also require a specific set of inquiry skills to bring meaning to their work. Gott, Duggan and Roberts (e.g. Roberts & Johnson, 2015) have been leaders and innovators in this space over the last two decades by highlighting that learners are not equipped to evaluate scientific evidence unless these skills are specifically and explicitly taught. This work provides insights into the fundamental ideas underpinning science inquiry skills and provides practical approaches to embedding them in practice to promote higher-order thinking along with scientific literacy.

Chapter 7: Assessment for student and teacher learning (Panizzon and Keast) signals a shift from some of the factors and considerations that underpin quality science education to the practices that enable this to be enacted—in this instance, in the area of assessment. This work has been influenced significantly by Black and Wiliam's research from the 1990s (e.g. 1998), along with Black's solo focus more recently (e.g. 2003, 2013), around assessment and its ability to be a tool for, of and as learning. This approach challenges traditional notions of assessment, particularly in science education, and the focus on teaching to a test, to envisage different types for different purposes. Informed by the voices of students and teachers, this work identifies a number of assessment strategies that contribute to quality science learning and teaching, and examines how they work, for what purpose and why.

Chapter 8: Pedagogical content knowledge (Loughran) explores an area often connected with more experienced teachers. Shulman (1986) introduced this idea over three decades ago, but the chapter author has pushed the boundaries of this construct in the area of science education for a number of years. In the context of this work, pedagogical content knowledge (PCK) refers to knowledge about the strategies and approaches that are effective in teaching the different aspects of science. This means moving beyond just teaching science in particular ways (e.g. hands-on learning, activities) because it seems to be the right thing to do, and instead engaging in science learning and teaching with a purpose in mind and articulating what that purpose is. Sometimes PCK can seem like an abstract construct. This chapter not only provides strategies to assist in developing science PCK, but pushes this body of knowledge further by making it accessible to future teachers of science.

Chapter 9: Transitioning into professional practice (Cooper and Smith) picks up where the previous chapter left off by turning the focus onto teacher development and professional growth. While this may not seem necessary when you are in the beginning stages of your career, as the analogy (learning to drive a car) that runs through this chapter (Speakman, 2012) suggests, these things take time and dedicated work, and will be dynamic over the course of your practice. Building on the research of Clandinin (1985) and Huberman (1989), at the heart of this work is teacher quality and thinking. However, in recognising your stage of career, practical suggestions around possible areas and approaches to professional learning are shared and conditions regarding how to position yourself as a professional learner are outlined. This includes using the AITSL (2011) and ASTA (2002) standards as a guide for paying attention to learning over your career as a teacher of science.

Chapter 10: Science education in informal settings (Reid and Liu) offers an opportunity to move outside the classroom and broaden your horizons about where science learning and teaching can take place. Notions of purposeful and authentic learning of science are a common thread running throughout this book, but there is something particularly valuable about learning in informal education settings and the value this brings. This work is situated within, and draws examples from, an outreach program at a zoo, but has relevance to other settings and spaces. The research of leaders in this field, such as Rennie (e.g. 2014), and Falk and Dierking (e.g. Falk & Dierking, 2010; Dierking et al., 2002), are significant influences, but this work pushes our thinking by developing the notion of science-scapes and what these can mean for enriching science education. It certainly signals a shift from the school excursion as a one-off, unconnected experience to thinking more deeply about the innovation that informal settings can bring to science learning and teaching.

Chapter 11: Science and STEM education (Corrigan and Lancaster) concludes the book by looking towards future directions for learning and teaching science. An important consideration is the role of science education, particularly in terms of how it contributes to the global imperatives for building STEM capabilities for future citizens. Science, technology, engineering and mathematics (STEM) are considered key areas for ensuring economic prosperity into the future, both as individual disciplines and as a collective. STEM education is pivotal to ensuring that our future thinkers are responsive and adaptable to emerging opportunities and challenges. Science is also important for the role it plays in developing expert thinking. This chapter explores how the education sector can address the goals for the twenty-first century, particularly critical thinking, creativity, collaboration and communication, through STEM education—even though STEM itself is not easily located within the school curriculum. Teachers and students can develop appropriate capabilities for their futures using both disciplinary expertise such as science and also STEM in interdisciplinary spaces. Examples from professional practice and learning are used to highlight the opportunities that can exist when these opportunities focus on addressing the needs of learners, including the relevance of such learning.

WHAT HAPPENS NEXT?

This is not your usual science education book. It will not provide you with a guide on how to write lesson plans or units of work. Nor will it provide you with a set of ready-to-go classroom activities or information that unpacks specific science content. It will, however, encourage you to think differently about science education, your role in it and what this means in a contemporary context. Equally, it will challenge you to question what science is and what this means for your subsequent science learning and teaching practice. A significant point of difference is that each chapter is not only steeped in research, which provides a strong evidence base *for* practice, but draws on case studies, vignettes and quotes to provide some direction about what quality science education looks like *in* practice. This book thinks big and encourages you to do the same as a teacher of science.

It is not our intention that you will necessarily read this book from cover to cover, though it has been structured in a logical sequence of practice to enable this approach. We do recognise, however, that you might instead choose to dip in and out of the chapters in ways that make sense to you and your own professional knowledge and growth over the course of your education career. We suspect that you will take away different ideas about science learning and teaching every time you delve in, based on your experiences and growing expertise. This book has been designed to provoke and challenge your thinking about and ways of practising science education; we don't imagine time will change this.

It is not feasible or useful to think that we can prepare you through your initial teacher education courses to know everything you will need to know in your career as a teacher. Not only is this next to impossible; it doesn't account for the dynamic nature of our education systems and how we think about what contributes to quality learning and teaching at different periods of time. One thing that is constant is change, so we need to prepare you to be ready and open-minded for that, as well as equipped to seek the information and insights you need at any given point.

Much of what we do in initial teacher education is plant seeds and hope that, with some time as well as the right nurturing and environment, they will blossom into quality learning and teaching practices. This book follows a similar logic.

Our approach has been to consider what key ideas and constructs you might need to be aware of as a teacher of science—not necessarily right now as a pre-service teacher, but certainly into the future.

DISCUSSION QUESTIONS

1.1 How do you currently view yourself as a teacher of science?
1.2 What factors do you think have influenced how you perceive yourself (or not) as a teacher of science?
1.3 Which chapters to you feel drawn to and why?
1.4 How do you imagine using research to inform your practice?
1.5 What aspirations to do you have for your students in learning science?

RECOMMENDED READING

Fitzgerald, A. (2012). *Science in Primary Schools: Examining the Practices of Teachers.* Rotterdam: Sense.

Hackling, M., Ramseger, J. & Hsiao-Lan, S. (eds) (2017). *Quality Teaching in Primary Science Education.* Dordrecht: Springer.

Loughran, J. (2010). *What Expert Teachers Do.* Sydney: Allen & Unwin.

REFERENCES

AITSL (2011). *Australian Professional Standards for Teachers.* Canberra: Education Services Australia. Retrieved from www.aitsl.edu.au/docs/default-source/apst-resources/australian_professional_standard_for_teachers_final.pdf.

Appleton, K. (2002). 'Science activities that work: Perceptions of primary school teachers'. *Research in Science Education*, 32(3), 393–410.

ASTA (2002). *National Professional Standards for Highly Accomplished Teachers of Science.* Australia: Australian Science Teachers Association and Monash University.

Black, P. (2003). 'The importance of everyday assessment'. In M.J. Atkin & J.E. Coffey (eds), *Everyday Assessment in the Science Classroom* (pp. 1–12). Arlington, VA: NSTA Press.

—— (2013). 'Pedagogy in theory and in practice: Formative and summative assessments in classrooms and in systems'. In D. Corrigan, R. Gunstone & A. Jones (eds), *Valuing Assessment in Science Education: Pedagogy, Curriculum, Policy.* Dordrecht: Springer, pp. 297–329.

Black, P. & Wiliam, D. (1998). 'Assessment and classroom learning'. *Assessment in Education*, 5(1), 7–74.

Clandinin, D. J. (1985). *Classroom Practice: Teacher Images in Action.* London: Falmer Press.

Claxton, G. (1990). *Teaching to Learn: A Direction for Education*. London: Cassell.
Corrigan, D., Buntting, C. & Jones, A. (2013). 'Rethinking the place of context in science education'. Paper presented at the annual conference of Australasian Science Education Research Association, Wellington, New Zealand.
Cowie, B. (2002). 'Re-viewing conceptual change through a formative assessment lens'. In R.K. Coll (ed.), *Science and Technology Education Research Papers* (pp. 164–79). Hamilton, New Zealand: Centre of Science and Technology Education Research, University of Waikato.
Dierking, L.D., Burtnyk, M.S., Büchner, K.S. & Falk, J.H. (2002). *Visitor Learning in Zoos and Aquariums: A Literature Review*. Annapolis, MD: Institute for Learning Innovation.
Duschl, R.A. (1983). 'The elementary level science methods course: Breeding ground of apprehension toward science? A case study'. *Journal of Research in Science Teaching*, 20(8), 745–54.
Falk J.H. & Dierking, L.D. (2010). 'The 95% solution: School is not where most Americans learn most of their science'. *American Scientist*, 98, 486–93.
Fitzgerald, A. (2010). 'Beliefs, knowledge and practices of effective primary school teachers'. Unpublished PhD thesis, Edith Cowan University.
—— (2012). *Science in Primary Schools: Examining the Practices of Effective Primary Science Teachers*. Rotterdam: Sense.
Goodrum, D., Hackling, M. & Rennie, L. (2001). *The Status and Quality of Teaching and Learning of Science in Australian Schools*. Canberra: Department of Education, Training and Youth Affairs.
Harlen, W. (1997). 'Primary teachers' understanding in science and its impact in the classroom'. *Research in Science Education*, 27(3), 323–37.
Huberman, A. (1989). 'On teachers' careers: Once over lightly, with a broad brush'. *International Journal of Educational Research*, 13, 347–61.
Kamler, B. & Thomson, P. (2006). *Pedagogies of Supervision: Helping Doctoral Students Write*. London: Routledge.
Lederman, N.G. (1992). 'Students' and teachers' conceptions about the nature of science: A review of the research'. *Journal of Research in Science Teaching*, 29, 331–59.
—— (1996). 'Syntax of nature of science within inquiry and science instruction'. In L.B. Flick & N.G. Lederman (eds), *Scientific Inquiry and Nature of Science: Implications for Teaching, Learning, and Teacher Education*. Dordrecht: Springer, pp. 301–17.
Mortimer, E.F. & Scott, P.H. (2003). *Making Meaning in Secondary Science Classrooms*. Maidenhead: Open University Press.
Pajares, M.F. (1992). 'Teachers' beliefs and educational research: Cleaning up a messy construct'. *Review of Educational Research*, 62(3), 307–22.

Prain, V. & Waldrip, B. (2006). 'An exploratory study of teachers' and students' use of multi-modal representations of concepts in primary science'. *International Journal of Science Education,* 28(15), 1843–66.

Reid, A. & Price, D. (eds) (2018). *The Australian Curriculum: Promises, Problems and Possibilities.* Canberra: ACSA.

Rennie, L. (2014). 'Learning science outside of school'. In N. Lederman & S. Abell (eds), *Handbook of Research on Science Education Vol. II.* New York: Routledge, pp. 120–44.

Roberts, R. & Johnson, P. (2015). 'Understanding the quality of data: A concept map for "the thinking behind the doing" in scientific practice'. *Curriculum Journal,* 26(3), 345–69.

Shulman, L.S. (1986). 'Those who understand: Knowledge growth in teaching'. *Educational Researcher,* 15(2), 4–14.

Speakman, J. (2012). 'What are we really trying to teach?' In S. Keast & J. Loughran (eds), *Exploring Scientific Literacy Through Cases of Teaching and Learning.* Melbourne: Monash Print Services, pp. 18–19.

Wenning, C.J. (2008). Dealing more effectively with alternative conceptions in science. *Journal of Physics Teacher Education Online,* 5(1), 11–19.

White, R.T. & Gunstone, R.F. (1992). *Probing Understanding.* London: Falmer Press.

CHAPTER 2
WHAT THE LEARNER BRINGS

KATHY SMITH *and* **ANGELA FITZGERALD**

LEARNING OUTCOMES
After reading this chapter, you should be able to:

- recognise that constructing shared understandings in science involves both students and teachers working together to share and clarify their own thinking and experiences
- identify that paying attention to the learner is critical to meaningful science education because both students and teachers bring a range of different ideas, experiences and beliefs
- be mindful that building shared understanding requires conditions for learning that value and invite a range of diverse perspectives and provide the opportunity for rich and interactive dialogue
- understand that when students and teachers work together in ways that value different perspectives, this potentially enhances the way knowledge is socially constructed.

> Learning is about building relationships with people—children and adults—and creating connections between ideas and the environment; not separating or isolating subjects, skills or people. (Daws, 2009)

Every day, in a range of situations, teachers make decisions about what matters for the learners in their science classrooms. They are required to create effective learning environments and build relationships that nurture and support the ongoing development of knowledge, skills and attitudes. Not only are teachers responsible for ensuring that every student has an equal opportunity to learn science in personally meaningful ways; they are also expected to provide conditions that develop each learner's capacity to deal independently with increasingly complex and abstract conceptual ideas and to use these ideas in their everyday life to assist them to make informed decisions. Science teaching is far more than simply delivering curriculum. There is no 'one size fits all' approach that works for every student, and this makes teaching problematic and at times challenging. Developing teaching expertise requires teachers to be willing to see themselves as learners: open to exploring new ideas, new ways of thinking and new ways of working. Actively engaging with students provides science teachers with opportunities to learn more about the complex interrelatedness between pedagogy, science knowledge and the contextual reality that defines their teaching. The development of such professional knowledge influences what each teacher pays attention to in their teaching: how they listen to, observe and take notice of their students' learning needs then how they choose to respond.

Expert teachers of science, in both primary and secondary school settings, value and invest time in learning more about their students and are able to effectively extend and enhance curriculum to enable each student to progress toward social and cognitive development.

As Hargreaves (1994) states, teachers shape and define curriculum based on what they think and believe, and what they do in the classroom ultimately shapes the kind of learning students experience. Expert teachers learn with their students: they value and actively attend to what the learner brings. Constructing shared understandings in science involves both students and teachers working together to share and clarify their own thinking and experiences, calling on any existing relevant knowledge they have to deepen understanding. Paying attention to the learner is therefore critical to meaningful science education because both students and teachers bring a range of different ideas, experiences and beliefs to science education, and these different perceptions influence learning. Building shared understanding requires conditions for learning that value and invite a range of diverse perspectives and provide opportunities for rich and interactive dialogue. When students and teachers work together in ways that value different perspectives, this potentially enhances the way knowledge is socially constructed.

This chapter explores the interplay between teachers and students as learners and the value of such interactions in developing shared understandings and personally relevant learning in science. It attempts to explore the importance of interactive dialogue as a way of deepening learning far beyond required curriculum outcomes. The next section illustrates the effectiveness of such interplay through a story that began with one student's question, which created a critical learning opportunity for both teachers and the student. In this scenario, the teachers and student, as learners, work together to establish meaning and deepen a mutual and shared understanding about an idea. At all times, teachers and student are learning together—they are at the centre of the learning experience. As the story illustrates, learning depends on what each learner brings and their willingness to engage in the process. As teachers, we believe that this story raises implications for quality science teaching, which will be discussed in a later section. We intend to raise your awareness about the importance of valuing and attending to each learner's personal experience and identity in science. This is equally important whether it be in a primary or secondary school setting.

GRAPPLING WITH A GOOD QUESTION

As lecturers, we often get emails from our pre-service teachers sharing something they may have thought of after a discussion in a workshop or seeking clarification on an idea or concept with which they might be struggling to connect. As learners ourselves, these questions are really valuable because they enable us to learn more about our students and often give us cause to reflect on shared conversations and experiences. Yet the email shared below, from one of our primary pre-service teachers, is a little different, in the sense that it has had a lasting impact because it caused us to question and more carefully examine our own science education understandings and practices—especially in relation to the ideas of and experiences with science that different learners bring to the classroom.

> Hello Kathy,
> I hope you don't mind me coming to you directly with a question I have about alternative conceptions and whether I am categorising something in my assignment correctly. I am one of your online students for [unit code], and I have really engaged with the first few weeks of the unit under your and Ange's guidance.
>
> Kathy, the science experience I am using as the basis of my assignment will be a unit of study I did in my middle years of primary school about my local environment. I was raised on the shore of Lake Mulwala, part of the Murray-Darling river system [in north-east Victoria, Australia]. As a very young child, I learned the

> Dreamtime story of the creation of the river system by a giant snake. I vividly recall taking this story to be the literal truth of the creation of the riverbed, right up until the unit of study I will be referencing, which was in Year 4. I think it is one of the reasons I remember this science experience so well—I was still a child who had fairies at the end of her garden at this time!
>
> My question is that, as this Aboriginal story had so shaped my understanding of my local environment up until that point, would it be right to label it as an alternative conception? I want to make sure that I am being sensitive to the Aboriginal tradition of storytelling and oral language. I believe there is significant merit in using these stories in the education setting and I certainly intend to do so when I become a teacher myself. I don't want it to seem as if I am diminishing the importance of these stories by categorising one's belief in them as an alternative conception.
>
> I welcome your opinion, Kathy!

Having just spent significant time in a workshop with this cohort of pre-service teachers discussing the range of experiences, backgrounds and understandings that learners bring to our classrooms and what this means for science learning, including the notion of alternative conceptions, this question initially took us by surprise. Oh no . . . that's not what we meant by alternative conceptions! But this surprise quickly shifted to reflection and contemplation. So what do we mean, then? Is this construct even a very useful way of making sense of the existing ideas that learners hold about science phenomena, or is it just confusing? How does the notion of alternative conceptions sit with different ways of knowing science? What followed was a number of lengthy discussions between the two of us, as well as with other science education colleagues, as we worked to untangle our own understandings of the ideas connected with science learning in this email. While it is difficult to capture all the things we talked about, the following three sections tease apart the essence of these conversations before revealing the response given to this email. The focus is on three areas: constructivism, alternative conceptions and different ways of knowing. When considering what the learner (students and teachers) brings to the classroom, these areas might be most relevant to you and your science teaching practice, in both primary and secondary school contexts. To begin, we will explore constructivism, a theoretical approach with which you may already be familiar, and consider what this lens might offer us in terms of enhancing our science learning and teaching. To assist in making sense of these three rather complex constructs—constructivism, alternative conceptions and different ways of knowing—we will draw on a number of examples to further illustrate the point being made from different perspectives.

What does constructivism offer us as teachers?

The learning experience shared in this email reminds us that science is all about the experiences we have with the world around us and the ways we piece together an explanation for a phenomenon that we encounter. These explanations draw on a range of past experiences, inputs from family and friends, belief systems, science knowledge and our own ways of processing information. We are required to draw on these different resources and inputs to make sense of our world all the time and at any age. This way of sense-making generally holds us in good stead and assists us to navigate our world, but our own constructed ways of understanding a phenomenon are not always consistent with scientifically accepted ways of understanding the same phenomenon. Consequently, when learners' explanations of different **science phenomena** have already been useful in making sense of a concept or situation, it can become difficult for them to view the concept or situation in question from a different perspective or through a different lens. This taps into a key challenge faced when teaching science in primary and secondary schools alike. Within a classroom of learners, each individual brings a vast array of different experiences and understandings that impact how they make sense of the science they are encountering at school. In parallel to this, teachers, as individuals, also interpret and then share their understandings of science as it is filtered through their own experiences. So what does this all mean for science learning and teaching? A useful theoretical lens that can assist us to make further sense of these complexities by further explaining the process of learning is constructivism.

science phenomena a science-based event or situation that is observed to exist or happen

Current understandings acknowledge the complexities that shape learning and teaching by exploring and identifying the **attributes**, behaviours and **learning outcomes** that are valued in the classroom. The theoretical construct of constructivism assumes that learners enter into any new learning situation with an array of previous experiences based on a mix of their cognition, feelings and skills that have been arranged to form individualised 'mini-theories' of understanding (Claxton, 1990). A main tenet of constructivism is that prior learning is used to make sense of new experiences and new information. Therefore, it is acknowledged that learners are not blank slates and that what they already believe and know matters (Levitt, 2001). Within this construct of constructivism, quality learning and teaching practices should provide students with the structures and environments the need to extend their understanding.

attributes a quality or feature that can be considered a characteristic of someone or something

learning outcomes what the students will be able to do at the end of the teaching sequence

This construct is further cultivated through **social constructivism**, a branch of constructivism that emphasises the influence of culture, context and social interactions on how learners develop and construct their understandings (Confrey, 1995; Driver et al., 1994). With learning viewed as both a cognitive and a social process, the context in which the learning occurs and the social influences that learners bring to the environment are crucial (Richardson, 1997). This suggests that quality learning and teaching practices—in all learning areas, not only science—should take into consideration the socially and culturally constructed understandings that learners bring to the classroom and provide ways of building upon or reshaping these understandings.

social constructivism
a sociological theory of learning that considers development to be socially situated and knowledge to be constructed through interactions with others

The previous paragraph was theoretically quite complex, so it is completely understandable if you are still not entirely sure what this might mean for student learning and your own teaching practice. To break this construct down further and illustrate what constructivism might look like in action, consider the following story.

> I still distinctly remember the pre-service teacher who bustled into my primary science education workshop brimming with excitement. She had a story to share. In the previous week, we had been discussing the ways in which children (and all of us, in fact) draw upon their previous experiences of the world to make sense of science phenomena. With this in her mind, the student thought she would ask her five-year-old son some science-related questions to gain a first-hand insight into this discussion. When she asked him the first question she had in mind, 'What are clouds made of?', she wasn't anticipating his reply. First, he looked at her with a rather

exasperated look that suggested the answer was obvious, then he sighed and responded in very definite terms, 'Mum, clouds are made of helium and fluff.'

—Angela Fitzgerald

Beyond this being an exceedingly cute response, this anecdote starts to get at the core of what constructivism 'looks' like. It shows how we call upon experiences we have had and things we know to bring together explanations of more abstract or unfamiliar parts of our world in ways that make sense to us. We are sure that you can imagine, like we do, that this child had been exposed to cotton wool or similar 'fluffy' materials at some point in his life and had experienced helium through the joy of receiving a helium balloon and observing its unique qualities. These two tangible experiences are then drawn together to explain the rather intangible notion of clouds and how they are formed. While this may seem simplistic and is more primary education focused in nature, these underpinning characteristics are no less relevant to how learners make sense of more complex science ideas and concepts, or how secondary-age learners may construct their understandings.

While constructivism can be viewed as a broad theory, sometimes seemingly without boundaries, there are several features of this construct that are extremely useful to us when teaching science in primary or secondary school settings. For us as teachers, it makes sense to draw on the tenets of constructivism when trying to understand how students make sense of their world. The following are the key aspects of this construct that might be particularly valuable to consider when planning and implementing relevant and engaging science learning experiences.

- Constructivism assumes that the learner is always cognitively active rather than passive and seeking to make meaning of what is taking place or being experienced in their environment.

- Constructions of meaning are conditioned and constrained by personal experiences. As each individual has unique interactions with their world, as a consequence all learners have a variety of perspectives they bring to the learning situation. With this in mind, constructivism frames learning as an interpretative process involving individuals' construction of meaning related to specific occurrences and phenomena (Watts & Bentley, 1994).
- Any new constructions are built in relation to previous experience and prior knowledge.

Constructivism helps us to think differently and more openly about our students and the ways in which they connect with science. It also helps us as teachers to think about our own ways of knowing and interpreting science phenomenon. This construct suggests that all learners bring something to the classroom with the challenge lying in how we, as teachers, value these understandings and integrate them into our science teaching practices.

In relation to the email we received, our student had constructed an understanding of her natural environment based on local Aboriginal perspectives. This next section will explore a specific aspect of constructivism known as alternative conceptions and what this means in terms of our science teaching, especially when making sense of what students bring to their science learning.

Alternative conception or not?

Constructivism emphasises that what is in the learner's mind matters because they are constructing their own meanings for the knowledge they acquire (White & Gunstone, 1992). A learner may communicate strong beliefs based on repeated experiences of a particular phenomenon. Sometimes these ideas and explanations may not be reflective of the ways in which scientists represent the world, but because the learner's ideas have already proven personally useful in making sense of their own world, these ideas are described as alternative conceptions. Hence alternative conceptions often result when new experiences are interpreted in light of prior experiences, and new understandings are grafted onto prior understandings (Wenning, 2008). The term **misconception** is also sometimes used synonymously with the phrase alternative conception, but in the literature they are given different meanings, with misconceptions carrying more negative connotations.

misconception a view or opinion that is incorrect based on flawed thinking or logic

Alternative conceptions of science phenomena are not unusual. They are very much part of the learning process across our lifetime. Our everyday experiences naturally have a significant impact on the formation of our ideas, but not all of the ideas that we develop are aligned with the most current evidence and scholarship in a particular scientific field. Sometimes things we have already learned are unhelpful in learning new concepts. This occurs when the new concept is inconsistent with previously learned material. Some concepts in different areas of science are simply difficult to grasp. They may be abstract, counter-intuitive or quite complex. Therefore, our understandings become inconsistent with scientific understandings. Teachers and students alike hold alternative conceptions about different scientific concepts (Burgoon, Heddle & Duran, 2010).

Before going much further, it might be useful to consider a classroom-based example where grappling with alternative conceptions of a scientific idea is evident and, unlike the previous cloud example, also complex. The example below is an excerpt from Fitzgerald (2012), and captures the discussion between four Year 7 students as they grapple with making sense of the properties of materials.

By way of context, the students had spent a number of lessons engaging with a range of materials and identifying their properties. To extend their understandings, the students were given a series of questions and were required to individually respond in ways that demonstrated they were able to apply their knowledge to a number of new scenarios. One particular question stands out, as it brings to the fore the students' thinking, in particular how they began to draw from prior experience to make sense of new information. In doing so, they not only revealed their conceptual understandings but also how they understood and were able to apply the nature of evidence to justify their explanation. The answers they convey in the following excerpt demonstrate how new explanations are often grafted onto prior understandings and that prior understandings may also be alternative conceptions. Therefore, alternative conceptions can underpin and/or be created as students construct and explain their understandings.

Cucumber on your eyes has been *scientifically proven* to reduce the wrinkles under them and make you look much younger. Is this true? Explain your answer.

Anna and Natalie indicated that this statement was true, while Evan stated that it was not true and Mark was not sure. Anna justified her belief that the statement was true by providing reasoning that linked the need for moisture in your skin to prevent wrinkles to the potential moisturising properties of cucumber.

'Yes, it is true because cucumber is moist and when you moisturise your skin it is stronger. If you have dry skin, if you fall over your skin will break and scratch. If it's moist it doesn't as much. Moisturiser also prevents wrinkles. So I suppose cucumber would too.'

> Natalie did not provide any reasoning to accompany her belief in the statement, other than stating, 'It is true because it is scientifically proven.' Evan did not agree with the statement and supported his belief by stating that 'cucumber's a vegetable and it's only blocking light from your eyes'. Mark did not agree or disagree with the statement, instead explaining, 'I do not know if this is true because I have not done a test.' (Fitzgerald, 2012, p. 92)

The question posed by the teacher was designed to challenge and extend student thinking. There are a number of conceptual ideas at work in this scenario, such as the observable properties of the cucumber and the relationship, if any, to the observable properties of the skin, the association of hydration with cell function and the relationship—if any—between cell hydration and skin appearance. Connecting all or some of these ideas in a meaningful and scientifically accurate way was particularly challenging for these students. The explanations provided demonstrate that as learners broaden their experiences, they potentially broaden their ideas and knowledge, and this means they often have increasingly more information to draw from to construct explanations. Yet determining what is relevant and useful knowledge can be a confusing process for the learner. For the teacher, assessing the source of confusion or the specific alternative conception that is at play may become a progressively complex task, particularly when students deal with increasingly complex problems.

Alternative conceptions can impact science learning in several ways, as shown in the above responses.

- Students generally are unaware that the knowledge they have may be inconsistent with scientific understandings.
- Alternative conceptions can be very entrenched in student thinking.

- Students interpret new experiences through these different understandings, thereby interfering with their ability to integrate scientifically accepted ways of understanding science phenomena.
- Alternative conceptions tend to be tenacious and resistant to instruction because learning entails replacing or radically reorganising student knowledge.
- Teachers need to notice how and why students are thinking in particular ways.

For learning to happen, **conceptual change** needs to occur. This puts teachers in the very challenging position of having to bring about significant conceptual change in their students' knowledge (Duit & Treagust, 1998).

conceptual change a shift in conceptual understanding to be more in accord with the scientific understanding of the situation

Misunderstanding, miscommunication, mis-education, and even a misapplication of well-established ideas, theories and principles can all lead to the formation of alternative conceptions (Wenning, 2008). Sometimes students can experience the same phenomenon and still draw different conclusions. Sometimes students might cling to false notions that are the result of a parent's, peer's or teacher's own alternative conceptions. In other instances, students might misapply what correct information they do possess, leading to what appear to be alternative conceptions. As teachers, we need to be acutely aware that alternative conceptions are not necessarily naïve viewpoints. As Clement, Brown and Zeitsman (1989) note, 'not all preconceptions are misconceptions'. In this instance, preconceptions are similar to alternative conceptions, as they are the conceptions held by the learner before a learning event. In the context of this chapter, not every different way of knowing or understanding science phenomena indicates the presence of an alternative conception. Some ideas that are inconsistent with scientific explanations are nothing more than students encountering difficulties in explaining new phenomena. The origin of any given alternative conception is essentially impossible to determine.

The reflection and dialogue below from Lisa, a primary school teacher, and her Year 3 and 4 students (Fitzgerald, 2012) provide further insights into how students grapple with making sense of abstract science ideas, such as day and night, and the impact this has on their learning; however, it goes on to illustrate the response a teacher might make in order to attend to the development of alternative conceptions and help to steer understanding towards scientifically accepted responses.

> As the students performed their role-plays depicting their understandings of how day and night occur, Lisa noted that a number of the groups had formed an alternative conception that the Moon was involved in this process. Lisa addressed this issue by again modelling how day and night occur using a torch (Sun), globe (Earth) and a tennis ball (Moon). She did explain that sometimes the Moon does block the Sun's light from reaching Earth, which is known as an eclipse. Lisa believed, in hindsight, that she should have left the Moon out of the role-play to lessen the conceptual confusion of the students.
>
> 'Unfortunately, I should have left the Moon right out of it because then they got the idea that the Moon was causing the day and the night. But I think by following that up at the end, talking about that idea of the eclipse rather than day and night really helped. [However] when I went around and was reading their responses to what causes day and night, [some of the students] still had that the Moon causes day and night.'

Lisa revisited the concept of day and night with the students a week later by showing an animation depicting how day and night occur. Lisa used this opportunity to reiterate this conceptual idea and readdress this alternative conception.

Teacher: What sorts of things did we learn last week about day and night? How does day and night happen? How do the Earth, Sun and Moon move together? Andrea?

Andrea: As the Earth moves around, one side is facing to the Moon and one side is facing the Sun. So the side that's facing the Sun, that's day and on the other side, that's night.

Teacher: Fantastic. Excellent. How does the Earth move? There were two ways that the Earth moved. Who can remember what one of those were? Ewan?

Ewan: It spins around on its own axis.

Teacher: It does. Very good. How else does the Earth move? Michael?

Michael: It moves around the Sun and the Moon moves around the Earth too.

Teacher: Fantastic, Michael. And as we spoke about last week, the Moon doesn't actually have anything to do with day and night. The Moon is still moving around the Earth. Sometimes the Moon gets in the way of the Sun's rays. Can anybody remember what that was called? It started with an E. Who's got a really good memory? Nathan?

Nathan: An eclipse.

Teacher: An eclipse. Excellent. But the Moon doesn't actually have anything to do with day and night. (Fitzgerald, 2012, pp. 103–4)

DIFFERENT WAYS OF KNOWING

Science is a traditional way of knowing that is reflective of Western culture. Understanding the world from a scientific perspective within the bounds of this culture means that ideas have been developed as a result of science discovery that includes rigorous interrogation and a search for evidence. Only after repeating this interrogation are explanations accepted and considered reliable and valid ways of explaining a particular phenomenon. However, not all ways of knowing are born from science. Science learning is complex as it needs to be heavily intertwined with each learner's reality, which is made up of factors such as their cultural background, linguistic background, learning needs, past experiences, parental expectations and their own interests, to name but a few. Explanations for phenomena developed through these different ways of knowing or interpreting life experiences can be cultivated in a variety of different ways, including through oral story telling or visual input. These approaches are usually steeped in long traditions and are very much experiential in nature, which could be considered forms of rigour and evidence.

Like the anecdote in the email from our student, the story below from Geoff, a secondary school teacher, draws on Aboriginal perspectives in relation to making sense of the world or, in this instance, a particular experience within a familiar environment. This reiterates the points being made above and further highlights the complexities inherent in the constructs of alternative conceptions (alternative to scientifically accepted understandings) and different ways of knowing (something that may have a scientific basis) as well as the importance of not using them interchangeably, which is explored in more depth following this story.

> I had already been teaching these 12 young Aboriginal men for a couple of years when this experience

occurred. By this time, they were 14 years old and had all been through a traditional initiation process with the exception of Kevin, who had spent numerous years in and out of this community and was the most enculturated of the group in terms of understanding Western ways. It was the dry season in the Top End of Australia, so I had decided we would spend the afternoon going on a bushwalk to take advantage of the milder conditions while we could.

As we walked through the limestone country and negotiated the low-lying scrub, the boys pointed out a number of important sites as well as collected and snacked on endless bush tucker. Always! In fact, on this particular walk, they managed to catch a huge goanna, which we carried with us for people back in the community to feast on later. It wasn't long after this that we were stopped in our tracks by a very unusual feature in the landscape—a large indent in the land that would have measured about 40 metres long, 10 metres wide and 6 metres deep. It was a very big sinkhole. But what was most surprising about this was the mini-rainforest that had taken up residence in the bottom, with lush ferns and intensely green growth looking extraordinarily different to the rest of our surrounds. We all stood around, in silence, peering in.

Through my eyes, I was making sense of this scene as most likely being caused by some land subsidence into an underground cave. But, given how significantly different this feature was in the landscape, I was surprised that this wasn't a sacred site and wondered if we were somewhere we shouldn't have been. I turned to the guys, 'What is this? Is this a special place?' I was

> reassured when someone piped up, 'Oh no Geoff. This is nothing.' But the response made me curious, 'So how do you think this came to be here?' I asked, gesturing at the sinkhole. My question was met with downward eyes and shuffling.
>
> I could sense in the pause that this was an embarrassed silence. Kevin was the one to finally speak up, which on thinking back was surprising given the amount of time he had spent in a Western school system. 'Bolung (the rainbow serpent prominent in the Dreamtime stories of the Aboriginal people of this region), he did it.' Then he looked away as he added, 'It's obvious.' I quickly sensed that any further questioning or discussion about this anomaly wasn't going to get me much further. In terms of the boys' understandings, what had happened here was indeed obvious and not worth another thought. So we turned on our heels and continued on our way. (Geoff, cited in Fitzgerald & Pfeiffer, 2017, p. 210)

The challenge for us as teachers lies in the differentiation of these two constructs: alternative conceptions in science and different ways of knowing. What is the difference? The notion of alternative conceptions focuses on drawing together everyday understandings of science concepts, but finding that they do not align with the accepted (Western-orientated) scientific views. For example, a primary school student might explain the experience of stirring a spoonful of sugar into a glass of water as the sugar disappearing rather than dissolving, while a secondary school student might explain the action of walking on the moon as being possible because of the lack of gravity, rather than there being less gravitational force than there is on Earth. These examples highlight an application of science understandings that are incomplete due to factors connected with the development of alternative conceptions as outlined above—often the abstract or complex nature of the concept. Different ways of knowing science are usually culturally constructed ways of understanding the world. They are not defined as alternative conceptions because these viewpoints are not alternative to Western scientific understandings; they are a completely different way of making sense of phenomena. In returning to our email, the example shared by our student of understanding the formation of a river through a story passed down over generations within her local Aboriginal community can be situated as a different way of interpreting a phenomenon. While this viewpoint is very different from Western scientific ways of explaining how a river system is formed, it is no less valuable or valid. What is presented here is not an alternative way of understanding geological formations; rather, it is an entirely different way of explaining an experience.

Framing a response

Our discussions and grappling with the issues related to the question posed in our student's email led to the following response.

> Hi,
>
> Thanks for your email. This is a really interesting question that you raise. My initial response is that Dreamtime stories remind us that school-based science presents very much a Western perspective in terms of the explanations that have been developed for everyday phenomena. The Dreamtime stories capture the Aboriginal perspective—that is, an alternative explanation rooted in traditional systems of beliefs that Aboriginal people have used for thousands of years to understand their environment and the changes and developments that take place,

particularly biophysical changes. These stories are an inherent and fundamental part of cultural identity and social integrity. While many of these stories provide an alternative explanation to Western science, the knowledge contained in these stories provides a wealth of wisdom and experience from direct observations shared mainly through stories over many generations.

I am thinking that there are many interesting inherent issues in your personal story that relate to science teaching and learning—obviously, the need to be culturally sensitive, the role of communication, empirically constructed knowledge and so on. However, I don't think the label of alternative conceptions has a place here or is a helpful way to think about or make sense of this experience. Your story captures two different ways of knowing and it is important to remember that the label 'alternative conception' was born within research around learning Western science, which is an empirically based way of knowing. Within this way of knowing, young children's thinking was examined to understand how children growing up in this tradition were making sense of their experiences and how their thinking developed. So we can see how a child's view within this particular tradition may move closer to the accepted scientific thinking that defines this perspective. Your story is bigger than that. Remember, it is possible for someone to sit comfortably with two ways of knowing (scientifically accepted and culturally constructed) without a hierarchical order being placed upon such knowledge.

This is a complex issue and you have generated some really interesting discussion for myself and my colleagues. I have cc'd Ange into this discussion as we spent some time talking about your email. I hope I have explained my thinking around this clearly, but I would encourage you to consider other more helpful ways of making sense of this experience. It is a really rich one to draw on for your future development as a science teacher.

Kind regards,
Kathy

This response was followed by our student's thoughts on what we proposed.

Hello Kathy,
Thank you for responding so thoughtfully (and quickly!) to my question. Your response has really helped me to order my thoughts around how to incorporate this into my assignment.

I had a feeling that looking at Aboriginal beliefs as alternative conceptions was not right, but I just couldn't quite find the words to articulate my thoughts. Your point about how the Dreamtime stories remind you that school based science is predominantly Western in its approach really made me stand back and think—that was when I realised that, without meaning to, I was prioritising my Western belief system ahead of the Aboriginal one.

I think this idea about there being two ways of knowing in my story, and finding a way to sit comfortably with this is at the crux of it for me. My husband is a medicinal chemist, and we have had many discussions

> on this topic over the years, with respect to how some of his colleagues reconcile their religious beliefs with their scientific work. We have had another really productive discussion on the 'two ways of knowing' topic this evening, and as usual he said lots of sensible, intelligent things. I think I will be asking him a lot of 'science' questions in the coming years!
>
> I think I might go to bed before my head explodes with all of these big, interesting ideas!
>
> Thank you again, Kathy.

As learners ourselves, this student's question raised an opportunity for some critical thinking, and this led not only to a deeper sense of our own thinking, but also an awareness of how our students may be grappling with the big ideas in our tutorials. The interplay, which involved discussion about different perspectives, enabled the development of a shared understanding, and together we all constructed meaning and clarified our ideas. The story demonstrates the importance of diverse perspectives, intellectual collaboration, purposeful questioning and critical reflection as essential elements of effective learning.

IMPLICATIONS FOR TEACHING

This story illustrates that learning is about changing thinking, and this is rarely a straightforward or predictable process. Learning often involves cognitive conflict, which can be quite a challenging, if not confronting, experience. Working through this process inevitably depends upon the learner's willingness to engage with the hard work of reconsidering what they know to determine whether such ideas are useful and stand up to interrogation. To do so requires conditions for learning where trust and respect are assured and where diversity is valued, allowing each individual to express learning in personally meaningful ways. As illustrated in the above exchange, a learner is far more likely to engage with such challenges when working within a safe learning environment where teachers and students are encouraged to explore different ways of thinking about and knowing the world. While curriculum is often framed in ways that suggest learning is linear, sequenced and predictable, in reality the process of learning is rarely so clear, direct, ordered or unproblematic. This has a number of implications for science teaching.

What do teachers need to pay attention to as they work to support students in being/becoming effective science learners? Working with teachers in a primary setting, Smith (2011) found that a number of fundamental actions emerged that became essential for teachers to ensure that each student was able to engage with science in a personally meaningful way. From our experiences, these actions are no different in a secondary school setting. Smith (2011) discovered that it was essential for teachers to be willing to confront their own ideas and beliefs about learning and the nature of science itself. Teachers needed to see themselves as learners with permission to ask questions and sit comfortably with uncertainty. In doing so, teachers implicitly model for their students important learning behaviours, such as questioning, **reflection** and **critical thinking**. Ultimately, this provides permission for both students and teachers to be learners together.

This thinking is captured very powerfully in the following quote, as experienced primary teacher Mary Howard shares her thinking about what matters for quality learning in her classroom. In this quote, Mary actively positions herself as a learner alongside her students, emphasising the role of questioning and sharing opportunities as ways of seeking understanding.

reflection time for contemplation and consideration

critical thinking involves the objective analysis and evaluation of an issue or event in order to form a judgement or reach a conclusion

> There is no doubt that we are on a continual learning path with our students, but it is often helpful to see the success of our strategies, to confirm the achievements we have made along the way and to celebrate the answers we discover. Now as I look back on my journey of the last five years, I can see that I have come from being a teacher in charge of a limited science curriculum for students to a confident risk-taker who allows students to lead the way within a cooperative learning environment in order to build up an evolving curriculum. I still don't have all the answers, but this hasn't prevented me from asking more and more questions. (Howard, 2011, p. 56)

It is also important for teachers to notice the conditions that enabled their own learning in science. Teachers need to use this information to shape the learning conditions they provide for their students. Expert teachers must be strategic in how they access student ideas and thinking, and work to identify each learner's individual skills, capabilities and life experiences. They must provide opportunities for learners to interact with their environment through social and collaborative processes, utilising discussion and diverse perspectives as opportunities to learn more about their students and the world around them. As shown in the story shared earlier, science teachers must provide safe and trusting learning environments in which diversity is encouraged and valued. By creating these learning conditions, teachers nurture curiosity and support their students to pose rich questions and be active learners. In the following quote, experienced primary teacher Tracey Adams reflects on the learning conditions she creates in her teaching and how they enable students to socially construct knowledge, value diversity and continually seek understanding.

The shift for me has been in reconsidering my practice in relation to the questions:

- What can I do to allow the students more input into their learning?
- How do I guide possible directions for students' learning while still allowing them room to explore?
- How can I challenge them to develop deeper understanding?

I have discovered that questioning is all important—particularly the type of questions I initially ask in a topic. These questions are paramount; if effective they invite a myriad of responses from my students. In return, I must be accepting of all their ideas and contributions without indicating what is right or wrong. All things are possible at this important stage.

Giving the students time to use their language to add to the discussion is vital. Any definitions or understandings presented are accepted. When I changed my teaching behaviour, such that I began to attend to these questions, I began to hear interesting comments from the students:

I like that definition.

I agree with you but what if we . . .?

Could it be . . .?

And from myself I began to hear:

Is there anything different that anyone would like to add?

That's what we think today. I wonder if we will think the same in a week? (Adams, 2011, p. 62)

Expert science teachers not only see themselves as learners and create conditions that enable quality learning; they also encourage students to take ownership of their learning. Yet while doing so they continue to monitor and check how students are making sense of information and experiences. Simple ideas are then seen as opportunities for rich learning, and when students are given time to develop deep learning they have opportunities to explore a range of possibilities and conceptual learning outcomes. Providing learning experiences that link learning to events and opportunities in the wider community, encouraging students to apply learning and take action in meaningful ways are all important for quality science learning. Most importantly, expert teachers of science are always working to keep learning 'alive', and they do this by making links across learning experiences throughout the year as opposed to positioning learning within and as being limited to individual units of work (Smith, 2011).

> My responsibility had shifted from organising the answers and all the work to fit into a neat little box, to supporting the learners in their journeys of discovery and at times this could be just as challenging for me as it was for them. I noticed they were beginning to make connections to their world, they were asking questions as much as finding answers and trying to work out how this learning connected to them and what they could do with this information. This was action that was helpful to their learning and their thinking. This was action that mattered. (Adams, 2011, p. 62)

Teachers as learners need to attend thoughtfully to the teaching approaches and processes they utilise each day so that these become objects of critical scrutiny. Teachers must consider how effectively these approaches challenge each learner's thinking and assumptions, value and build upon diverse perspectives, enable ownership of learning and require learners to articulate how they are linking and making sense of ideas and experiences. This is challenging work yet, as the following quote from experienced teacher Ann France indicates, when a teacher recognises this work as part of their own learning journey, then expertise in teaching continues to develop.

> Sadly I think in my first years of teaching I was probably guilty of missing opportunities to witness such fantastic discussions. Like many teachers, I may have contributed to a belief that there is only one answer that will make the teacher happy. Students learn to become compliant and respond accordingly. However, I think it is crucial that I encourage my students to be independent learners, to take charge of their own destinies, their own thinking and their own opinions. To do this, I had to take a big step back and share intellectual ownership with my students and allow them to explore, hypothesise and find things out for themselves, even prove themselves wrong. (France, 2011, p. 103)

CONCLUSION

Paying attention to what the learner—whether they be a student or a teacher, in a primary or secondary school setting—brings is central to effective and quality learning in science. But it requires an acknowledgement that learning is an active process. We all bring a range of ideas and explanations based upon our experiences and interactions, and we use this information to make sense of situations and phenomena we experience in our daily lives. This has significant implications for science learning and teaching. It requires teachers themselves to be active learners of science through building their own expertise and knowledge so that they can, in turn, create effective conditions for their students to undertake meaningful learning in science. While, as a pre-service teacher, giving consideration to what a learner brings to a science classroom may not seem like a pressing concern as you grapple with issues connected with planning, assessment and classroom management, we urge you to think differently. By opening your mind to the diverse ways in which learners think about and experience science, you are opening up a number of opportunities for understanding how to better support learning in this area.

DISCUSSION QUESTIONS

2.1 As a teacher in a primary or secondary school setting, how might you value and incorporate understandings such as these into your science lessons?

2.2 How might you build upon what might be considered non-scientific ways of making sense of a phenomenon to enhance more scientific ways of viewing the world?

2.3 This chapter uncovers the importance of teachers noticing the conditions that enabled their own learning in science. In what ways might you be able to move forward with your own thinking?

2.4 Consider an area of science that you have some well-formed ideas about e.g. how gravity works. What strategies might you use to find out if your own ideas are consistent with the scientific view of this idea?

2.5 In the vignette of night and day presented above, what suggestions would you have for the teacher to improve her role play in ways that may reduce students' alternative conceptions?

FURTHER READING

Gunstone, R.F. (1995). 'Constructivist learning and the teaching of science'. In B. Hand and V. Prain (eds), *Teaching and Learning in Science: The Constructivists Classroom*. Sydney: Harcourt Brace, pp. 3–20.

Mitchell, I. (n.d.). 'Children's alternative conceptions in science'. Retrieved from: http://monash.edu/science-education/wp-content/uploads/2017/07/altern concepts.pdf.

Monash University Science Education Research Group (SERG) & Victorian Department of Education and Early Childhood Development (DEECD) (n.d.).

Science Continuum F–10. Retrieved from http://monash.edu/science-education/2015/resources/the-science-continuum-f-10.

REFERENCES

Adams, T. (2011). 'Action that matters'. In J. Loughran, K. Smith & A. Berry (eds), *Scientific Literacy Under the Microscope: A Whole School Approach to Science Teaching and Learning*. Rotterdam: Sense, pp. 59–65.

Burgoon, J.N., Heddle, M.L. & Duran, E.D. (2010). 'Re-examining the similarities between teacher and student conceptions about physical science'. *Journal of Science Teacher Education,* 21(7), 859–72.

Claxton, G. (1990). *Teaching to Learn: A Direction for Education.* London: Cassell.

Clement, J., Brown, D. & Zeitsman, A. (1989). 'Not all preconceptions are misconceptions'. *International Journal of Science Education,* 11, 554–65.

Confrey, J. (1995). 'How compatible are radical constructivism, sociocultural approaches, and social constructivism?' In L. Steffe & J. Gale (eds), *Constructivism in Education*. Hillsdale, NJ: Lawrence Erlbaum, pp. 185–228.

Daws, J.E. (2009). 'Teachers and students as co-learners: Possibilities and problems'. *The Journal of Educational Enquiry*, 1(6), 110–26.

Driver, R., Asoko, H., Leach, J., Mortimer, E. & Scott, P. (1994). 'Constructing scientific knowledge in the classroom'. *Educational Researcher,* 23(7), 5–12.

Duit, R. & Treagust, D. (1998). 'Learning in science: From behaviourism towards social constructivism and beyond'. In B. Fraser & K. Tobin (eds), *International Handbook of Science Education.* Dordrecht: Kluwer, pp. 3–25.

Fitzgerald, A. (2012). *Science in Primary Schools: Examining the Practices of Effective Primary Science Teachers.* Rotterdam: Sense.

Fitzgerald, A. & Pfeiffer, L. (2017). 'Celebrating Australia's diversity in science education'. In G. Woolcotts & R. Whannell (eds), *Senior Secondary Science: Theory and Practice*. Melbourne: Cambridge University Press, pp. 191–213.

France, A. (2011). 'Speaking about scientific literacy'. In J. Loughran, K. Smith & A. Berry (eds), *Scientific Literacy Under the Microscope: A Whole School Approach to Science Teaching and Learning*. Rotterdam: Sense, pp. 101–9.

Hargreaves, L. (1994). *Changing Teachers, Changing Times: Teachers' Work and Culture in the Post-modern Age*. New York: Teachers' College Press.

Howard, M. (2011). 'You don't have to have all the answers'. In J. Loughran, K. Smith & A. Berry (eds), *Scientific Literacy Under the Microscope: A Whole School Approach to Science Teaching and Learning*. Rotterdam: Sense, pp. 47–57.

Levitt, K.E. (2001). 'An analysis of elementary teachers' beliefs regarding the teaching and learning of science'. *Science Education,* 86(1), 1–22.

Richardson, V. (1997). 'Constructivist teaching and teacher education: Theory and practice'. In V. Richardson (ed.), *Constructivist Teacher Education: Building New Understandings.* London: Falmer Press, pp. 3–14.

Smith, K. (2011). 'Learning from teacher thinking: An insight into the pedagogical complexities of scientific literacy'. In J. Loughran, K. Smith & A. Berry (eds), *Scientific Literacy Under the Microscope: A Whole School Approach to Science Teaching and Learning.* Rotterdam: Sense, pp. 25–36.

Watts, D.M. & Bentley, D. (1994). 'Humanising and feminising school science: Reviving anthropomorphic and animistic thinking in constructivist science education'. *International Journal of Science Education,* 16(1), 83–98.

Wenning, C.J. (2008). 'Dealing more effectively with alternative conceptions in science'. *Journal of Physics Teacher Education Online,* 5(1), 11–19.

White, R.T. & Gunstone, R.F. (1992). *Probing Understanding.* London: Falmer Press.

CHAPTER 3

THE ROLE OF CONTEXT AND RELEVANCE

DEBORAH CORRIGAN *and* **ANGELA FITZGERALD**

LEARNING OUTCOMES

After reading this chapter, you should be able to:

- recognise that context and relevance have an important role in science education if authentic and meaningful learning is to take place
- acknowledge the range of factors that contribute to and impact on a teacher's ability to emphasis context and relevance in their practice
- draw upon and apply different frameworks to enrich science learning experiences
- value the role of teachers' reflections in making the nuances of science education practice, and the components that shape it, visible.

One of the challenges that primary and secondary teachers face in teaching science is finding the right **context** in which to position science ideas and concepts to best support the learning of their students, in the

context the circumstances that surround the environment, settings and/or experiences that influence learning

authentic something genuine or original

fluency the ability to move seamlessly between different realms such as scientific ideas and everyday ideas

disposition the inclination of sensitivity and ability to act

discernment shrewd decision-making that requires both fluency and disposition

competence the mastery of fluency, disposition and discernment

sense that it is relevant, **authentic** and meaningful in terms of how the students experience and think about science in their daily lives. In meeting this challenge, Corrigan, Buntting and Jones (2013) developed a framework to help maintain the focus on science learning experiences that are contextually situated and relevant to the learner. This framework, which they called 'ways of knowing', purposefully interprets knowing as a verb rather than simply a noun. This is to emphasise that knowing is much more than just having a body of factual knowledge to access; it is an active and dynamic process that is inclusive of actions such as **fluency**, **disposition**, **discernment** and **competence** that need to be developed. The relationships between these actions is represented in Figure 3.1.

This active way of thinking about knowing plays out on the four dimensions located on two continua as represented in Figure 3.2. On both axes there are two dimensions located at either end of a continuum. In terms of this framework, science knowledge can range from individual concepts to big ideas (see the vertical or *y*-axis), but it can also be represented by—and representative of—simple

Figure 3.1 Actions necessary for ways of knowing science in contexts

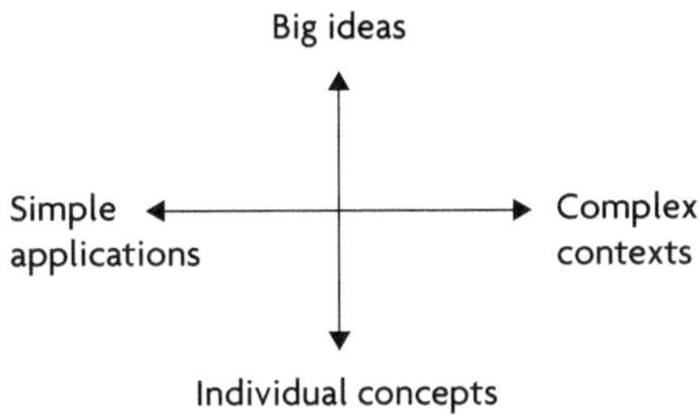

Figure 3.2 The 'ways of knowing' framework

applications through to complex contexts (see the horizontal or x-axis). Each continuum is explained in more detail below.

individual concepts discrete conceptual ideas

big ideas capture the notion of grouping together individual concepts that are linked in some way

FROM INDIVIDUAL CONCEPTS TO BIG IDEAS

The notion of **individual concepts** is less problematic than that of **big ideas**, as it represents all the discrete conceptual ideas of science. Examples include the idea of force being a push or a pull, living things not existing in isolation, melting as the change of a solid to a liquid, and the Earth moving around the sun. These individual concepts are often listed as bullet points in science curricula, and are usually the focus of summative assessment tasks such as tests and examinations. They are discrete and often definitional, and accessing student knowledge about them is relatively easy. While the concept of 'big ideas' has been used in a number of ways, it generally involves grouping together a number of individual concepts that link together in some way.

Big ideas are often represented by theories and models, such as the conservation of mass or energy, relativity, the cellular basis of life, evolution and geological time. An alternate view of a big idea is the use of a focus question to guide learning in a particular field—for example, 'Why is water a special substance?' or 'How do we interact with the atmosphere?' Other examples of big ideas centre around themes as seen in the conceptual strand maps contained within the AAAS's (2001, 2007) *Atlas of Scientific Literacy*, which is based on headings such as Force and Motion, Living Things and Nature of Science. Generally, what is common to all of these types of big ideas is that they are based around conceptual understandings.

FROM SIMPLE APPLICATIONS TO COMPLEX CONTEXTS

Simple applications, for the purposes of this chapter, are defined as examples of how science might be used where there is a straightforward one-to-one relationship between the application and the science concept. Such applications are often 'added on' to illustrate concepts, and are peppered through most school science textbooks. For example, pictures of vapour clouds produced from 'dry ice' at a concert are used to illustrate a solid converting to a gas in presenting a model of solids, liquids and gases; speed cameras use the idea of distance versus time measurements; and enzymes make chemical reactions more likely and are used in manufacturing sweets, washing powders and for tenderising meat.

Complex contexts represent combinations of concepts related in multi-faceted ways. Examples include socio-scientific issues such as genetic modification and climate change. Such complex contexts are often interdisciplinary in nature and require sophisticated ways of conceptualisation best suited to team thinking. They also tend not to sit just within the discipline of science, but need to bring together many ways and types of thinking and acting in order to be understood. To construct a more complex context therefore requires expertise that is often beyond teachers' own experiences and time that many teachers may not have. Such

simple applications refer to examples of how science might be used or applied in practice

complex contexts a collection of concepts that are connected in multi-faceted and dynamic ways

programs are difficult for teachers to conceptualise and construct, so it is probably no surprise that such teaching experiences are not common in science classrooms.

The 'ways of knowing' framework provides teachers with an intuitive way to ensure the science conceptual learning taking place in their classroom meets the needs of learners by being contextually situated and relevant. However, as teachers, we know that there are also a variety of different factors and conditions of which we need to be aware if we are to effectively address all aspects of the context in which teachers and students find themselves positioned. This chapter unpacks a number of contextual considerations for science education through sharing two cases written by practising teachers: Gillian (primary school) and Diane (secondary school). Both cases reflect on **critical moments** that caused each teacher to think differently about science learning and teaching. In focusing on the role of context and relevance in science learning, we make sense of each case with these constructs in mind. For this analysis, we are moving away from the *ways of knowing* to take a more holistic perspective that encompasses a variety of factors.

critical moments not necessarily big events, but rather significant moments that cause pause for thought about teaching practice

For this chapter, we are considering the following seven components as they emerged from the two cases to assist in unravelling the contextual features at play:

- the conditions for engaging learning experiences in science
- a need to know
- an understanding that it is not just about science
- teaching considerations
- the reality of school structures
- management of expectation and risk
- the type of accountability involved.

Like the 'ways of knowing' framework, this approach offers another way to consider the roles of context and relevance in engaging students in meaningful and authentic science education.

We start with Case Study 3.1, which was written by Gillian Sullivan, a primary school teacher. Primary school teachers bring a significant amount of pedagogical expertise to what they do in the classroom (Smith & Fitzgerald, 2013), something that Gillian's story brings to life. While most people would acknowledge gaps in their science content knowledge (Tytler, 2007), they more than compensate for this by a willingness and ability to develop engaging learning experiences in science for the benefit of their students. The trick for any teacher, however, lies in judging the length of that string (a reference to the case study title): where and how far do you take the learning? This is the challenge with which Gillian grapples throughout her reflection as she shares with us the rewards and constraints she faces as she fosters a science learning environment rich in context and relevance for her students.

CASE STUDY 3.1: How long is a piece of string?

Gillian Sullivan, St Peter's Primary School, East Bentleigh

Teachers know science is exactly the kind of subject that is a conduit to great, engaged learning. The trouble is *how long is a piece of string?*

All the 'ducks were lined up' for a fabulous year! A move south to Grade 3 [working with students aged eight and nine years], that wonderful age when children are still thoroughly engaged, a crack team of colleagues with which to teach and a shift to the new Australian Curriculum which was finally going to mandate the teaching of more science in the primary classroom. So with the 'perfect class', virtually free of classroom and playground behavioural issues and their teacher being enrolled in the STaL [Science Teaching and Learning] program [a professional learning opportunity jointly implemented by Catholic Education Melbourne and Monash University], the year was set to be an opportunity to really teach one of my passions: science.

On those initial planning days, at the beginning of the year, we planned with gusto. The team had taught this unit, Living and Non-Living Things, the previous year, and it was pretty straightforward, with plenty of resources available in the school and on the internet. It would involve the type of classification and sorting work that typified junior school science and we were once again throwing around the usual ideas of how we might end the unit having grown some flowering plants timed for Mother's Day!

Yet pretty quickly came the realisation that this was a *short* term. Easter was around the corner. The crack Grade 3 team worked hard and smart; if we were to get everything on the 'term planner' ticked off, my personal STaL-inspired science rich notion might have to be revised.

By March, the unit was progressing but some corners had to be cut due to time pressures. The day had come when we were to plant our seeds, as a final conclusion to the unit. Melbourne was enjoying an 'Indian Summer' [a period of unseasonable warm and dry weather] and so in the cool of the morning, class 3G headed to the veggie patch to plant our rocket seeds.

My first offence

I was about to commit my first offence in the name of science! This was the time usually used for the precious morning literacy block.

We'd be quick, I told myself! Twenty minutes and we'd be done and dusted. We commenced our planting in an orderly way, revisiting all the scientific vocabulary learned from the classroom and giving consideration to factors that would sustain the seeds' growth. I stepped well back to allow the children to undertake the planting. *And that's when the magic started to happen!*

They broke naturally into small cooperative groups. How lovely, I thought. No one came to me with the usual whinge.

'I don't have a group.'

I started roaming in the background and eavesdropping. Mmmm, this was really sounding good. They were reading planting instructions, measuring the depth of soil, noticing the soil type and the way the mulch was helping to retain the moisture. As I looked more closely, I noticed the level of inclusiveness. They were sharing ideas and using inclusive body language. Wait, they were even sharing their tools sensibly! (Always a really sensible idea when it comes to spades and pitchforks!)

Oh my gosh, they were even correctly using the scientific language we had used in the classroom!

So something had gone on after all! A glance at the watch and it was 10.00 am already! Definitely time to move inside.

Over there in the corner was a pile of 20 or 30 sunflower plants, which had graced our veggie patch for most of the summer. They now lay in a jumbled pile after being rooted out at the end of the season. There seemed to be an explosion of excitement from the class as they made a beeline to the triffid-like flower heads, which seemed to exemplify everything we had discussed in class. The cycle of life and death was again repeated as the plants ensured their continued survival by each bearing a multitude of seeds. It was almost a Pentecostal moment! But oh no, it was now 10.15 am!

As I agonised over whether to return to the classroom or not, it was too late—the class was fully engaged! They were investigating, exploring and discussing the symmetry they observed in the sunflowers. Some were even posing mathematical questions and attempting solutions to their own wonderings by estimating the number of seeds in a single sunflower. They then ran to the other spent flower heads in the garden, and started to collect the seeds and discuss conditions for their optimal germination. They would go on to establish a seed bank and then sow them to grow new plants in the school gardens the following year.

And here lies the dilemma. The unit was meant to be drawing to a close, not cranking up. NAPLAN was looming, the term planner was set in concrete!

What to do?

Should we allow the students time to build a really meaningful learning experience?

Should I be a maverick and step out of line with the term planner and the other three grades?

Was it worth rocking the boat? After all, I was new to teaching the year level.

Should I proceed in the knowledge that what we were doing today was actually on the planner for next term anyway, so I could make up time then?

Could I really consider skimping on NAPLAN preparations in core subjects like English and mathematics for the sake of some additional science?

Clearly, engagement is gold and the children had virtually self-directed and designed their own inquiry, which would tick every box with regard to engaged learning.

All teachers know that ultimately these opportunities are exactly what a quality learning environment endeavours to provide.

My second offence

I now committed my second offence in the name of science!

Yes, the children and I indulged ourselves over the next two weeks and pursued our science project.

Predictably, it was deemed a great success by all of us! Of course, this could have been considered foolish.

What if the NAPLAN results came back and revealed my cohort was the weakest?

Once, this question would not have even rated in my thinking, but in today's world of accountability, it does sit there at the back of a teacher's mind.

Primary teachers largely have the 'luxury' of calling the shots with their allocation of time and are skilled at judging the attention span of their children to maximise their learning throughout the day. Yet, like our secondary teacher counterparts, we must be completely accountable for teaching our specified number of topics and outcomes.

Thus my burning challenge remains striking the balance between meeting the demands of my mandated term planner and creating exciting opportunities for my students to seriously explore a topic.

Source: Sullivan (2013), pp. 58–9.

CONDITIONS FOR ENGAGING LEARNING EXPERIENCES IN SCIENCE

A phrase used by Gillian near the start of her story—'all the ducks were lined up'—suggests the importance of creating the right conditions to enable positive and engaging learning experiences in science (Van Zee & Roberts, 2001). One of these conditions might be passion or interest in the topic or learning area—something that Gillian brings to her practice—which may result in a certain amount of personal motivation and enthusiasm (Ziman, 2000). But this is perhaps not as important as a willingness and openness to learn alongside the students. This emerges through Gillian's story as a critical component of the experience, with a clear sense of her stepping back, putting aside her concerns and letting the students direct their own learning.

Through Gillian's story, we are given insights into the value of enabling student-led exploration rather than scripting every aspect of a learning experience. She could have introduced the sunflowers in a more structured way, but the dialogue and interest may not have been as genuine and enthusiastic as it was when driven by the students' natural curiosity (Mortimer & Scott, 2003). This experience makes us think about teachers providing various science learning opportunities, but knowing when to stand back and relinquish control to allow the students to guide their way through experience. Gillian's skill in knowing when or when not to intervene in the learning process demonstrates the development of her pedagogical knowledge (Shulman, 1987).

A need to know

Linking learning activities with an authentic task, as Gillian has—in this case, in the school garden—is an important part of creating opportunities for students to directly use and apply their knowledge of science (Buxton, 2006). Part of creating such opportunities is about appropriately matching the activities and the task so the teacher generates a genuine need to know in the students (Osborne & Wittrock, 1983), which leads to a reason for wanting to work collaboratively and talk about their discoveries. In particular, there needs to be a circumstance for wanting to use scientific language if there is to be a shift from using more everyday terms to describe scientific phenomena. The most impressive part of this story is that Gillian's students were able to effectively integrate scientific terms into their own talk about what was happening as they explored the garden (Mercer, 2008). Such an interchange of language types provides evidence of the students' fluency, in terms of an ability to move between **scientific interpretations** and **everyday interpretations** and understandings of their world. Gillian also reflected on how time seemed to pass very quickly during these moments of intense student engagement with the task at hand.

scientific interpretations relate to making sense of an event using science as a basis

everyday interpretations relate to making sense of an event using everyday experiences as a basis

When the context is relevant and authentic, then time flies and rich learning is happening without even being really aware of it. This scenario reminds us of the notion of flow in learning (Csikszentmihalyi, 2014), which suggests that when you are doing something you find interesting and relevant, there is a tendency to get totally immersed in it.

It's not just about science

While this story provides insights into the impact and contribution of this experience on the development of the students' literacy and numeracy skills in contextually meaningful ways, it also seems that science is providing a context for the authentic development of several other skills and learning areas. It is important to remember that engaging in science learning experiences is not only about science, but provides an opportunity to learn other knowledge, skills and different ways of examining situations. Examples in Gillian's story include creative thinking as seen in a science context, and questioning.

TEACHING CONSIDERATIONS

Gillian's reflection is a case in point for the important roles that adaptability and responsiveness play in teaching if these moments are to be captured and capitalised on. For example, the original plan of planting seeds in the garden may have steered away from flowers for Mother's Day, and then to managing the constraints faced, but the integrity of the intended learning objectives was maintained. While primary school teachers are particularly skilled at being able to adapt to the circumstances with which they are faced, the school structures within which they work provide them with some different freedoms to their secondary school colleagues. This is not to suggest that secondary teachers do not and cannot achieve similar things. It just requires a different level of out-of-box thinking.

A point to note here—one that is relevant to both primary and secondary science teachers—is the role of the work associated with doing activities in learning opportunities such as the one we have been exploring here. With the end-point of this experience a product for consumption, which will be a special experience for the students—sharing and eating something they have grown—care must be taken not to lose sight of the science learning happening. It is easy to fall into the trap of focusing on the fun, hands-on nature of an activity and to downplay the learning components (Appleton, 2002). The end-point might be meaningful and relevant, but the **cognitive** and **affective** journey to this point should be too.

cognitive relates to using thinking processes and skills to make sense of an event

affective relates to using moods, feelings and attitudes to make sense of an event

The reality of school structures

School structures have the potential to undermine the responsiveness of any teacher in relation to how they engage with their own practice and the directions they would like to take student learning (Lam, 2005). Regardless of how frustrating these structural impediments may be they are just part and parcel of being a teacher; essentially punches we have to adapt to and roll with. The challenge Gillian encountered was being responsive to her students' interests while dealing with the realities of implementing the curriculum in ways that were accountable to her colleagues, school, parental expectations and national testing. Working with and within (and possibly around) the school structures to create productive learning opportunities that may not compromise the science rich aspects of an experience is a risk for any teacher. The fact that Gillian reflected on her actions as an "offence" is a challenge to the way that many believe that learning and teaching in primary schools should be structured, which suggests that there is still some way to go in shifting thinking and practice in ways that enable risks to be undertaken and managed. It seems that never too far from a teacher's mind are the conditions and systems they work within and the tensions this creates for fostering quality learning.

Managing expectation and risk

A really interesting point, in relation to this particular chapter, that caused Gillian some tension is the notion of stepping out of line with the learning and teaching happening in the other classes as well as what has been prescribed on the term planner. While continuity across a year level is an important consideration, it is worth thinking about the difference between conceptual continuity and contextual continuity. It is possible for different classes to cover the same ideas but in ways that draw on the different interests and curiosities of the individual classes. When consideration is given to both conceptual and contextual continuity in teaching approaches it opens up different approaches and relationships with term planning, which is not so linear and predictive. Reflecting on the conditions that were created by Gillian through this learning experience uncovers a number of factors that may have contributed to the quality science learning taking place in this experience. For example, conditions in this experience have included creating opportunities to talk about what is happening and why, allowing student-directed experiences, experiential opportunities, shared and collaborative ways of working, and an investment of significant time to explore. All of these conditions represent a degree of risk for teachers as they may counter what is expected. Consequently Gillian will need to manage this risk as she works with her year level colleagues and within the accountability structures of the school.

What type of accountability?

The concerns voiced here by Gillian are very real for all teachers—primary and secondary, especially those working with the year levels that are required to participate in the National Assessment Plan—Literacy and Numeracy (NAPLAN). If we flip these concerns around, though, an interesting thought bubble arises. Perhaps contextually rich and conceptually integrated approaches to learning better prepare students for these types of assessment because they have engaged with ideas and skills in authentic, in-depth and meaningful ways, which may suggest that they have a better chance of applying their understandings to different and more complex contexts.

However, in these situations, the teacher's role shifts from 'training' to reproduce knowledge (as in 'teaching to the test') to explicitly linking how knowledge can be applied in a number of settings (facilitating multiple conceptual links across contexts). Regardless of how you repackage it, the rise of accountability in education systems in a variety of forms, with a simultaneous decrease in teachers' accountability to their students (which is often the motivation for becoming a teacher—that is, making a difference for the students), represents a significant barrier to fostering the conditions for quality learning for teachers to negotiate. It is an interesting irony to contemplate: teachers know what they need to do to enhance student learning, but the school structures with and within which they work can be an inhibiting factor.

In your own practice, it will be important to reflect on these tensions, as Gillian has done, and work out where the opportunities lie for you and your students. A sense of balance in terms of the type of accountability with which a teacher engages is important to maintain.

******

Case Study 3.2 is written by Diane Lodge, a secondary school science teacher. Secondary school teachers bring strong conceptual understandings along with in-depth connections to a particular learning area to what they do in the classroom. This is acknowledged

abstract conceptual area refers to a concept that is not easily represented in a physical or concrete way

in Diane's growing awareness that just because she is interested in a particular area of science—in this case, chemistry—and implicitly understands how it links to other science ideas and concepts, that doesn't mean her students share that knowledge. This is the challenge with which she grapples throughout her reflection as she reveals her thinking about and modification of her practice to better enable her students to connect in more authentic and relevant ways to an **abstract conceptual area**.

CASE STUDY 3.2: Seeing the link, making the link

Diane Lodge, St Aloysius College, North Melbourne

I often think that students today miss out on so much by the way of learning experiences outside of the classroom. While they have access to infinite sources of information via a range of electronic technologies, they don't seem to know what to do with what they find, or how to apply it appropriately to understand new situations. For instance, I have found that students do not seem to be able to link what they learn in their science classes with what is happening in the world around them.

This issue of lack of 'linking' knowledge became very apparent to me during Year 11 chemistry class this year. Part of the curriculum requires students to look at a timeline of the development of different ideas that have led to our current understanding of the atom.

The approach that I regularly use is to give the students a prepared timeline, then go through the timeline events with them and try to stimulate their interest with attention-grabbing, but not necessarily scientific, titbits of information—for example, did you know that Antoine Lavoisier not only developed the 'Law of Conservation of Mass' but was also an unpopular noble person who was guillotined in the French Revolution? Usually my fascinating facts are enough to stimulate student interest and discussion. However, this year was different. I was met with blank faces when I expected a buzz of interest. Still, I ploughed on.

'Has anyone heard of Pierre and Marie Curie?' I asked.

'Yes Miss,' answered one student. 'They were on an episode of *The Simpsons*.'

Students then spent the next few minutes trying to work out exactly which episode. Things did not improve when we moved on to Neils Bohr and I asked whether anyone had seen a recent SBS documentary that looked at the rivalry between Bohr and another scientist, Max Planck. Another student pointed out, 'We're not all rich Miss. We don't have Foxtel.' At this point, alarm bells were ringing loudly in my head. How could I expect these students to understand the driving forces behind scientific development if they did not understand the context in which the development was taking place? Thankfully, the lesson ended soon afterwards.

I made my way back to the staff room bemoaning the fact to all and sundry that students have no understanding of history. Over lunchtime discussions with other teachers, I realised that one of the problems for the students was that I had presented the material in isolation and was expecting them to make links that perhaps were not as obvious as I had anticipated.

In the days before our next class, I decided to take a different approach. I set up a new timeline that had nothing to do with chemistry. I made a timeline of my life and alongside it I put a timeline of what was happening in the world at different points, including scientific discoveries, births and deaths of world figures, wars, political events, sporting events and historical events, such as the fall of the Berlin Wall. My idea was that the students could use this approach with the atomic theory timeline that I had given them, and construct alongside it a matching timeline of major world events. I hoped that by doing this activity, they would be able to see some of the triggers for why particular developments in atomic theory had occurred.

The following week, I was away on Year 12 retreat. I left the timeline task for my students to complete while I was gone. When I returned, I was amazed (and very pleased!) to find the students had been really keen to do the task. In fact, there was almost a competition to see who could find the greatest number of events that linked to the development of atomic theory.

Each class member managed to submit a timeline that met the criteria of matching world events to aspects of scientific development. One of my concerns with the task was that I would not be able to discern the level of understanding that students had developed about the link when I was assessing their work. Surprisingly, this was not the case. The students who had developed an in-depth understanding of what was required were able to identify the difference between events of world significance that needed to be included in the timeline and those that were only significant to a particular region or country of the world, and therefore of less importance in contributing to overall scientific development.

On reflection, I have gained some valuable insights through this experience into the relationship between teaching and learning science. Students do have an interest in real-world science, but it doesn't necessarily come from the same place as my interest in science knowledge. It is up to me to tap into students' interests, and I need to find motivating ways to do this. The links that I have always imagined existing between different subject areas are not obvious to everyone. I cannot assume that just because a student studies history, they will link what they have done in that subject with what we are doing in science. Now that I am aware of this, I need to include some way of enabling these links to be discovered and built on through my teaching. I don't claim that this activity has led to students suddenly being more aware of the place that science has in the world, but I do believe that the students enjoyed the activity and in the process became aware that a lot of scientific development has been driven by societal or political forces. What we, as teachers, need to remember is that our job is not to change the students to our way of thinking—of course we see the world differently—but to change our way of teaching so it accesses and builds from students' interests.

Source: Lodge (2009), pp. 56-7.

CONDITIONS FOR ENGAGING LEARNING EXPERIENCES IN SCIENCE

Diane's experience highlights that it could appear counter to 'good' education if students cannot link what they learn in school with what they experience in life. While the catch-cry is often that 'science is irrelevant', it is perhaps more accurately portrayed as science *education* is irrelevant. Stuckey and colleagues (2013) have explored the characteristics of relevant science education and come up with the following definition:

> Science learning becomes relevant education whenever learning will have (positive) consequences for the student's life. Positive consequences can include:
> - fulfilling actual needs related to a student's personal interest or educational demands (of which learners are aware), as well as
> - the anticipation of future needs (of which students are not necessarily aware).
>
> Relevance in science education covers both intrinsic and extrinsic components. The intrinsic dimensions encompass student's interests and motives; the

> extrinsic dimension covers ethically justified expectations of one's personal environment and the society in which they operate and live. (Stuckey et al., 2013, pp. 19–20)

Given this definition, it would seem that Diane was faced with a daunting task to find a way to make the historical account of the development of the atom, a required element of the senior chemistry curriculum, relevant to her students both intrinsically and extrinsically. For example, she was needing to consider how this curriculum requirement might contribute to her students' personal interests now and in the future (**intrinsic**) as well how it might contribute to their social competence and their career choices (**extrinsic**). The challenges in this task are evidenced in Diane's recognition that she needed to seriously 'rethink' the conditions under which such a task would become relevant. Her contention of *linking knowledge* seems to be just the beginning.

intrinsic refers to something that is innate or coming from within

extrinsic refers to something that is operating from outside or externally

Science education should provide students with experiences of the work of scientists that are as authentic as possible. Included within this work is the interrogation of scientific work that has already been conducted. Diane's timeline could be seen as fulfilling this purpose. If this is the case, then it is important to make such purposes explicit to students. Interrogating previous work involves critical analysis of the work that has been carried out. In this instance, some lines of examination that Diane could have pursued with her students included identifying whether there were flaws in the work of Lavoisier and querying why Niels Bohr and Max Planck were arguing. For example, was it to do with the work in which they were engaged or with who was funding their work? All of these contextual elements play an important role in helping to provide students with a more authentic, and ultimately more engaging, science learning experience.

A need to know

Generating a need to know around the historical development of ideas connected to the model of the atom may seem an impossible task. Diane counters this through recognising that the point of this task is to generate an understanding of the developmental nature of scientific understandings and knowledge. In this instance, there is a particular focus on the way science generates useful explanations and models that can then be used to explain other events or predict future events. Diane's example of significant events in her own lifetime provided students with an appreciation of more recent history and what might be considered significant. In many ways, this approach provided a necessary bridge for the students when linking an older period of time with chemistry-based ideas. However, considering how these ideas can have a positive influence on students' lives provides another piece in this learning puzzle, particularly when students do not all have the same interests and connections with science. The ability within this task to make your own decisions about what constitute significant events at the time helps to meet this need. The next challenge is linking to the present day in meaningful ways, which may be achieved by using tangible examples such as the quest for dominance in nuclear weaponry. As Diane reflects, 'Students do have an interest in real world science but it doesn't necessarily come from the same place as my interest in science knowledge.'

Underlying this task is establishing (convincingly) why a student would need to know about the model of the atom. If this is to happen, then being aware of our purposes for needing to know things that seem so removed from everyday life is critical. In this instance, an understanding of the atom and how it behaves is fundamentally important in pursuing chemistry. For example, chemical reactions are all about the rearrangement of particles (atoms), and building new materials is all about knowing how atoms behave in terms of bonding, attractions, emission or consumption of energy when they rearrange. Developing strong conceptual understandings of structure is fundamentally important in chemistry and the model of the atom is an integral part of this. In connecting with Diane's experience, establishing the reason for needing to know is an essential element that is often missing from the teaching of more abstract scientific ideas.

It's not just about science

As touched upon above, it is difficult to consider how a concept like the model of the atom will have positive consequences for a student's life. This is where an examination of the social conditions that existed around and promoted the development of a particular science idea becomes important. In other words, what was going on at a given point in time that meant scientists needed to come up with such ideas? Diane's approach was to instil in her students a recognition that science was also influenced by society and the events happening at the time, not just by science ideas and innovations in isolation. The development of a model of the atom was incredibly important to society at the time because the race for the development of the atomic bomb as a weapon in the world wars was underway, which was certainly driving scientific thinking and actions in a particular direction. Other contemporary examples could include the imperative to find cures for cancers or address climate change. It is all about what is happening in society at the time and understanding how the scientific world has decided to respond.

TEACHING CONSIDERATIONS

Within Diane's story, her students moved from an individual science idea to a deeper understanding of how this idea exists in a social context and becomes relevant to their own lived experiences. Additionally, they made their own connections to the science being discussed through pop culture references such as the Simpsons and Foxtel. These insights do raise the question: relevance to whom? From a teacher's perspective, within the learning and teaching of science, relevance to the student is paramount, but this is not the only type of relevance that exists and to which students are required to respond (Fensham, 2000). Table 3.1 highlights different stakeholders who may be referenced by teachers, and the types of relevance they consider important.

Diane's case draws on a number of these types of relevance. For example, she has attended to the *wish-they-knew science* as the

Table 3.1 Stakeholders and relevance

Types of relevance	Who decides what is relevant?
Wish-they-knew science	Academic scientists, education officials, many science teachers
Need-to-know science	The general public, who have faced and resolved real-life problems/decisions related to science and technology
Functional science	People in science-based occupations
Enticed-to-know science	The media and internet sites
Have-cause-to-know science	Experts who have interacted with the general public on real-life issues
Personal-curiosity science	Students themselves
Science-as-culture	Interpreters of culture, who can determine what aspects of science comprise features of local, national and global culture

historical development of the atom as seen by academic scientists, education officials and many science teachers, viewing it as fundamentally important for any student of chemistry. While this may indeed be the case, what is often missing is articulation of why these groups of experts see this as relevant. In the reiteration of the task, Diane has attempted to address both *enticed-to-know science* and *science-as-culture*, as students were required to work out the cultural or social context that existed at the different times in which this scientific work was located. The reliance on media and the reporting and recording of the history at these times is important. Additionally, this task tried to address *personal-curiosity science*, as students needed to make decisions about what they thought were important cultural and social events. It could be argued that before Diane modified the task, the types of relevance addressed may only have included *wish-they-knew science* and *science-as-culture*. When developing science learning experiences, it is important to consider what type of relevance is in operation, as quite commonly it is the *wish-they-knew science* that dominates.

The reality of school structures

Teachers work within highly structured systems, which present constraints to learning and teaching experiences. For example, as drawn from the case above, the 'following week [Diane] was away on Year 12 retreat. [She] left the timeline task for [her] students to complete while [she] was gone.' Being a secondary school teacher is not only about teaching your specialist class, but taking a holistic perspective to the education of young people. This requires participation in a range of activities outside the classroom and subject areas that are essential to a balanced education, but not mandated in the curriculum.

Secondary schools have been designed for students to study subjects as separate entities rather than in the highly integrated way that ideas and concepts tend to exist in real-world contexts. This approach maintains a focus on detailed facts over deep understandings of big ideas. This approach is unlikely to change, particularly in the senior years of secondary school, which means that secondary school teachers like Diane need to consider what can be done within these constraints to better support contextually rich science learning.

Managing expectations and risk

Diane points out in her reflection that the 'links that [she has] always imagined exist between different subject areas are not clear and obvious for everyone'. This highlights the impact that assumptions can have on closing down learning and learning opportunities. Adopting reflective techniques can enable teachers to crack open their assumptions and look at the relationship between learning and teaching from a different, more **student-centred perspective**.

At any stage of your career as a teacher, it is important to talk with other staff, work with colleagues, participate in reflective discussion and continually engage in professional learning. These practices form an important base for any teacher to be equipped with the necessary skills and attributes to manage what is expected of them and their students, as well as manage the risks to their professional work, which includes accountability from different stakeholders and the nature of working with young people who are still finding their individuality. Diane sums this up by stating that, 'On reflection, [she has] gained some valuable insights through this experience into the relationship between teaching and learning science.'

student-centred perspective suggests a point of view that is driven by the students and their world-views and interests

What type of accountability?

Diana's case highlights that what teachers need to remember is that their job is not to change the students to their way of thinking, but to change their way of *teaching* so that it accesses and builds from students' interests. While it is easy to fundamentally believe in this statement and understand that it contributes to quality learning, the challenge is enacting it in practice. In terms of accountability, the challenge for teachers is shifting thinking in terms of relevance and context from a teacher perspective to a learner perspective. As Diane has shared, being mindful of this is a significant step in the right direction. In pushing practice further, it might be about taking steps towards making the connections between science and science education more apparent and in ways that will bring science learning to life.

CONCLUSION

It is part of human nature to want to make sense of our world and the experiences we encounter. We are essentially curious and questioning beings. It is perhaps no surprise, therefore, that context and relevance have a critical role to play in science education, particularly if the learning taking place is to be meaningful and of assistance as we negotiate our way through different aspects of our lives. While this is understood implicitly by primary and secondary teachers of science, it remains a challenge to achieve in the classroom. This chapter, through the reflections of Gillian (primary teacher) and Diane (secondary teacher), acknowledges that there are a number of factors at play, such as school structures, expectations and accountability. However, rather than fixating on what could be perceived as barriers or constraints, the challenges that are teased out here are reimagined as possibilities to further enhance science learning and teaching. A key message that emerges a number of times through the cases and the subsequent analysis is the importance of ensuring that the focus is on the students and their learning needs. It is easy to teach science in ways that are of relevance to us as teachers—areas of personal interest, curriculum concerns and expectations of colleagues and families. This, however, is counter-intuitive to achieving quality science education outcomes for students. Instead, as teachers, we need to move outside our own comfort zones, as Gillian and Diane both did, and focus on our students—to examine what is capturing their interest and what learning path they are wandering down. If we can bring a greater awareness to teaching science in ways that are contextually embedded and relevant, then we are moving in the right direction to ensure that students' experiences of science are meaningful and that they matter to them.

DISCUSSION QUESTIONS

3.1 In reflecting on your classroom experiences (like Gillian and Diane did through their stories), what experiences or incidents can you identify that might have shaped your thinking about the important of context and relevance to science learning? What impact did they have and why?

3.2 How might you use the frames shared in this chapter to bring relevance and authenticity to your science teaching practices?

3.3 Based on your own context, what are some of the resources (human and/or material) upon which you might be able to draw to ensure that science learning is happening in ways that are meaningful and personally relevant?

3.4 Primary school teachers tend to create many opportunities for hands-on experiences in their approach to science education, but sometimes the learning can be lost among the excitement. What strategies might you draw upon to ensure that learning activities are purposefully and meaningfully linked to science as well as being engaging and interesting?

3.5 A key challenge when working in secondary school settings is the potential lack of flexibility that school structures can impose on the learning environment (e.g. amount of time per class, timetabling, etc.). What strategies might you put in place to work within these structures while still ensuring that your students experience science in context and as relevant?

FURTHER READING

Corrigan, D., Buntting C., Jones, A. & Loughran, J. (2018). *Navigating the Changing Landscape of Formal and Informal Science Learning Opportunities*. Dordrecht: Springer.

REFERENCES

AAAS (2001). *Atlas of Science Literacy. Volume 1.* Washington, DC: AAAS.

—— (2007). *Atlas of Science Literacy. Volume 2.* Washington, DC: AAAS.

Appleton, K. (2002). 'School activities that work: Perceptions of primary school teachers'. *Research in Science Education*, 32(3), 393–410.

Buxton, C.A. (2006). 'Creating contextually authentic science in a "low-performing" urban elementary school'. *Journal of Research in Science Teaching*, 43, 695–721.

Corrigan, D., Buntting, C. & Jones, A. (2013). 'Rethinking the place of context in science education'. Paper presented at the annual conference of Australasian Science Education Research Association, Wellington.

Csikszentmihalyi, M. (2014). 'The concept of flow'. In *Flow and the Foundations of Positive Psychology.* Dordrecht: Springer, pp. 239–63.

Fensham, P.J. (2000). 'Providing suitable content in the "science for all" curriculum'. In R. Millar, J. Leach & J. Osborne (eds), *Improving Science Education: The Contribution of Research.* Birmingham: Open University Press, pp. 147–64.

Lam, Y.L.J. (2005). 'School organizational structures: Effects on teacher and student learning'. *Journal of Educational Administration*, 43(4), 387–401.

Lodge, D. (2009). 'Seeing the link, making the link'. In A. Berry & S. Keast (eds), *Looking into Practice: Cases of Science Teachers' Professional Growth, Volume 1.* Melbourne: Monash Print Services, pp. 56–7.

Mercer, N. (2008). 'The seeds of time: Why classroom dialogue needs a temporal analysis'. *Journal of Learning Sciences,* 17(1), 33–59.

Mortimer, E.F., & Scott, P.H. (2003). *Making Meaning in Secondary Science Classrooms.* Maidenhead: Open University Press.

Osborne, R.J & Wittrock, M.C. (1983). 'Learning science: A generative process'. *Science Education,* 67(4), 489–508.

Shulman, L.S. (1987). 'Knowledge and teaching: Foundations of the new reform'. *Harvard Educational Review,* 57(1), 1–22.

Smith, K. & Fitzgerald, A. (2013). 'Making sense of primary science'. In A. Fitzgerald (ed.), *Learning and Teaching Primary Science.* Melbourne: Cambridge University Press, pp. 1–13.

Stuckey, M., Hofstein, A., Mamlok-Naaman, R. and Eilks, I. (2013). 'The meaning of "relevance" in science education and its implication for the science curriculum'. *Studies in Science Education,* 49(1), 1–34.

Sullivan, G. (2013). 'How long is a piece of string?' In S. Keast, G. Lancaster, J. Loughran & D. Panizzon (eds), *Teaching and Learning Science Through Cases.* Melbourne: Monash Print Services, pp. 58–9.

Tytler, R. (2007). *Re-imagining Science Education: Engaging Students in Science for Australia's Future.* Melbourne: Australian Council for Educational Research.

Van Zee, E.H. & Roberts, D. (2001). 'Using pedagogical inquiries as a basis for learning to teach: Prospective teachers' reflections upon positive science learning experiences'. *Science Education,* 85, 733–57.

Ziman, J. (2000). *Real Science: What It Is, and What It Means.* New York: Cambridge University Press.

Part II

Understanding science

Grappling with what science actually is and how this can be enacted in schools in ways that are authentic and not replicating a dated view of science is one of the great challenges we face as teachers of science. If one of the key aims of science education is to prepare students to be scientifically literate citizens—able to make informed decisions and choices based on evidence—then understanding how science is understood and practised matters. This means ensuring we are enacting science learning experiences in our classrooms that move beyond seeing science as a fixed set of facts governed by a particular methodology by presenting a more contemporary and holistic perspective.

Part 2 teases out three key areas for consideration:

1 *science curriculum*—unpacking what curriculum is and how it can be implemented in ways to support quality learning and teaching in science
2 *nature of science (NoS)*—articulating what NoS represents and making sense of how it can effectively and authentically be drawn upon to enhance science education
3 *science investigations*—thinking differently about what constitutes the investigative process in science and using pedagogies that will enhance learning of and from this process.

This section of the book is valuable because it will support you to better understand how, through engagement with the science curriculum documents and a shift in focus to embedding the nature of science in science learning and teaching, you will be more able to enact science investigative practices that are authentic and relevant in the classroom.

CHAPTER 4

THE SCIENCE CURRICULUM

DEBORAH CORRIGAN *and* **KAREN MARANGIO**

LEARNING OUTCOMES

After reading this chapter, you should be able to:

- articulate a personal understanding of what curriculum means
- understand the purposes of curriculum
- demonstrate an appreciation for the meaning of curriculum from the perspective of different users
- demonstrate your understanding of how to implement curriculum
- articulate an understanding of why curriculum is important.

This chapter explores the ideas, influences and purposes of curriculum, highlighting its multiple, complex and contested place within society. The focus is on understanding curriculum policy (the **intended curriculum**)—in this case, the purpose and structure of the Australian science curriculum, which acts as a reference point for schools to create and teach their own school science curriculum (the **implemented curriculum**). This chapter emphasises the importance of understanding curriculum policy and the teacher's integral role as a curriculum developer to provide meaningful experiences for their students (the **experienced (realised) curriculum**).

intended curriculum usually relates to the curriculum policy documents or the written curriculum; can be a broad and flexible framework such as the Australian Curriculum

implemented curriculum usually relates to the school's or teacher's curriculum—what they construct, plan and teach

experienced (realised) curriculum the students' experience and what they learn, including the hidden and null messages; relates to the knowledge and perspectives the student brings, their readiness and ability to learn, and their interaction with the curriculum and what they learn (intended by the teacher or not). It encourages you to articulate your own personal understanding of what 'curriculum' means to you and why it is important

WHAT IS CURRICULUM?

> When I started MTeach, it surprised me that even though others in my class were teaching science at the same year level on placement, we were not teaching the same thing and most were not using a textbook. Science is different in different schools, even though these schools were 'using' the same curriculum. [On placement] most of us were teaching Victorian curriculum, a couple taught interstate (ACT and SACE curricula) and IB (curriculum). Cool. (James, first-year MTeach pre-service teacher)

Education is a field full of jargon, and as a beginning teacher 'curriculum' is one of those jargon terms that we never really understand. This is hardly surprising as, on trawling through academic research literature, there does not appear to be much agreement on what 'curriculum' means. For example, views on school curriculum include:

- 'an interrelated set of plans and experiences that a student undertakes under the guidance of the school' (March & Willis, 1995, p. 10)
- 'Some claim that a curriculum is the content or objectives for which schools hold students accountable. Others claim that a curriculum is the set of instructional strategies teachers plan to use' (Posner, 1995, p. 5)
- 'the curriculum is always, in every society, a reflection of what the people think, feel, believe, and do' (Smith, Stanley & Shores, 1950, p. 3)
- 'Curriculum is both a noun and a verb, encompassing all the experiences that occur within the school or other learning context—what is intended and what actually becomes reality' (Pinar & Grumet, 1976).

In *The Curriculum* (1918), the first textbook published on the subject, John Franklin Bobbitt said that curriculum, as an idea, had its roots in the Latin word for racecourse (*currere*—to run a race). He explained the curriculum as the course of deeds and experiences through which children become the adults they should be for success in adult society. Furthermore, the curriculum encompasses the entire scope of formative deeds and experiences occurring in and out of school, and not only experiences occurring in school, both experiences that are unplanned and undirected, and experiences intentionally directed for the purposeful formation of adult members of society.

While curriculum is not a new idea, one of the reasons why curriculum has no universal meaning is that understanding curriculum is a highly personal activity—it will mean different things to each individual. So even if there is a well-articulated curriculum, the messages that such a curriculum sends will be different for every individual, based on their personal beliefs and experiences. From these descriptions, it is clear that curriculum encompasses all those experiences children have as they progress to adulthood. However, we very rarely consider curriculum in these terms, and instead tend to focus on the school or university curriculum. Hence curriculum is usually associated with a learning institution.

In school education, 'curriculum' and 'syllabus' are often used synonymously. While curriculum includes the subjects studied in a school or college and details of what each subject includes, a syllabus (from the Latin word *syllabus* for 'list') is usually a plan showing the subjects or books to be studied in a particular course, especially for a course that leads to an exam (Cambridge English Dictionary). A curriculum is generally more encompassing than a syllabus. For instance, the Australian science curriculum presents a broad and flexible framework—a big picture. It is not a syllabus and does not give detailed units of work or lesson-by-lesson plans of what to teach, how to teach it and in which lesson. This essential task is for teachers to undertake within their schools. This helps explain why, as James noted at the start of this chapter, science 'looks' different in different schools even though the same curriculum is being used. As such, the Australian curriculum recognises that context matters, including the role of the teacher in creating a science curriculum to suit their students at their school. A teacher's personal understanding of the curriculum and the ways in which it could be implemented, including how they value the purpose and importance of the science curriculum, takes centre stage. Implementing the curriculum is a highly personal activity.

In this chapter, we use curriculum in a number of ways, including dividing the school curriculum into the *intended, implemented and experienced (realised)* aspects of curriculum. Our main focus is on curriculum policy (in this case, the Australian curriculum) and we will call this the *intended curriculum*. This is a starting point for teachers to construct, plan and teach their *implemented curriculum*. Finally, the *experienced (realised) curriculum* is the student's experience of learning science. This relates to what students think, what they learn and what messages they are receiving about learning science in school. The manner in which the implemented curriculum is enacted by the teacher will send a range of messages to their students, including those about science and learning, beyond the intentions of the lesson(s) (whether intended or not), and some messages may leave a lasting impression into the future. Refer to Figure 4.1 for an advanced organiser of these components of curriculum. In this advanced organiser, you will see key questions that will

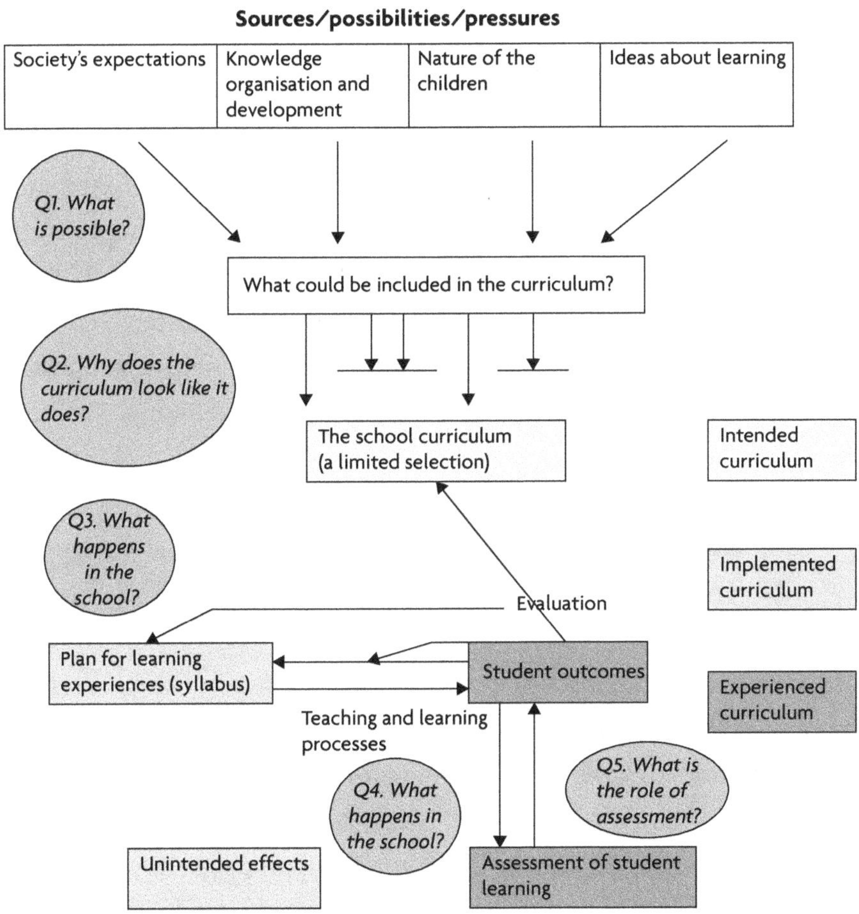

Figure 4.1 An advanced organiser for the intended, implemented and experienced curriculum

guide the discussion in this chapter. In addition, you will see links between aspects of the diagram, such as student outcomes linked to the sub-heading related to experienced curriculum. As you read through the rest of this chapter, think about what curriculum means to you. Think about your own experience as a school student learning science and your recent experiences in schools. What is your overall vision for science being taught in schools—why should your students study science? What do you want them to learn and why? What should be the purpose of the curriculum, and can there be a range of purposes for science in the school curriculum?

THE PURPOSES OF CURRICULUM

Consider what you think is the purpose of curriculum. The purpose(s) of curriculum, within which the science curriculum sits, is often contested by a range of people with different perspectives. According to social theorists—Young (1971) and Waring (1979) in Britain, and Gintis (1972) and Apple (1979) in the United States—school systems (including curriculum) are established to achieve a range of purposes that relate to societal needs and demands, such as:

- the reproduction of knowledge, skills and expertise
- the development of people with new skills and knowledge
- the conservation of traditional knowledge, values and culture.

For individuals, school systems can mean different things, such as the opportunity for social mobility that is associated with securing highly skilled and valued employment (and avoiding low-paid work), or the reinforcement of existing advantages, such as retaining advantaged social status. For some individuals, school systems may contribute little or nothing to the advantages or disadvantages that already exist. This brings into question the value and relevance of school curriculum to cater for everyone's needs.

The purpose of the science curriculum

Consider the range of purposes curriculum may have in the context of learning science. What should a science curriculum in the twenty-first century emphasise? Should science remain part of the compulsory years of schooling, and if so, why? Science's place within the curriculum adds another layer of complexity, with more perspectives and motivations to discuss and debate its purpose. Layton (1994) suggests that the science curriculum, in any given period, is the result of four different pressures on the education system or the sub-system of it that is concerned with a subject like science:

- political (environment of the country)
- economic (environment of the country)
- cultural transmission (passing on those cultural traditions of the society)
- subject maintenance (remaining 'true' to the importance, status and structure of the subject).

However, when considering these four pressures it is important to remember *whose* politics, economics, culture and subject are being identified, as this raises interesting tensions with regard to shaping any curriculum. Since the 1950s, two broad societal demands have been made of science education: the demand for *specialist science workers* and the demand for a *scientific literate citizenry*.

As Fensham (1985) points out, the demand for specialist workers is important for the maintenance and development of the social, economic and defence needs of countries. This demand has tended to dominate science education, and society has paid a price for this dominance. With the production of future scientists as a priority for science education, there has been a generation of 'knowledge of worth' (Fensham, 1985, p. 417), which consists of the consolidation of specific conceptual content in each of the sciences. The learning of such knowledge can take so much time that more exciting, contemporary and socially useful aspects of science are omitted or overlooked—aspects that would support the development of scientific literate citizens. Balancing the demand for specialist science workers with the demand for science-informed and interested citizens, whether or not they continue studying science at school/ university, requires serious consideration.

Reiss (2007) focuses specifically on science curriculum. His view of science curriculum includes all those things that take place during the time students are at school (ages five to eighteen). He suggests that *autonomy* (the ability to decide things for yourself), *wellbeing* (mainly personal wellbeing) and *justice* (usually social justice, as the right treatment of others and the fair distribution of resources or opportunities) are possible aims of education. The science curriculum, he suggests, could aim for the following:

- a supply of future scientists (though he echoes Fensham's sentiments about the cost to society)
- scientific literacy (helping tomorrow's adults understand scientific issues)
- individual benefit (there should be some benefit to the student)
- democracy (that science education should promote democracy in order to develop citizenship)
- social justice (as described above)
- criticality—thinking that is 'rigorous, logical, open-minded and penetrating, and may have elements of **reflective scepticism**' (Reiss, 2007, p. 2).

reflective scepticism
looking back on the questioning you have undertaken; relates to a general attitude of doubt or questioning of knowledge, opinions, beliefs presented as facts or claims that are made. Scepticism in science is often seen in the practice of subjecting beliefs and claims to scientific investigation in order to support their reliability

The possibilities for what could be included in any science curriculum are endless. It is therefore necessary to rationalise and compromise regarding what is included in the science curriculum.

MOVING FROM INTENDED TO IMPLEMENTED TO EXPERIENCED CURRICULUM

At this stage, you should recognise that different pressures contribute to the curriculum. In Case Study 4.1, Louise Van Corler (2013) shares with us an experience that she had creating a curriculum for her Year 6 students. The advanced organiser in Figure 4.1 (earlier in the chapter) highlights this filtering process as the curriculum moves through the *intended curriculum* (curriculum policy documents), the *implemented* (by teachers) *curriculum* and the *experienced or attained* (by students) *curriculum*. Louise's case emphasises the critical role of the teacher as she draws on a number of sources (teachers and students) to implement the Australian curriculum (the *intended curriculum*) and create powerful learning experiences for her students (the *experienced curriculum*) and herself.

CASE STUDY 4.1: Confusion is a good thing—even in teachers

Louise Van Corler

When faced with new subject requirements, sometimes the only way through is to embrace confusion, trust your colleagues and draw upon your students' input.

Planning: From blackness to dense fog

Eighty-nine students, all with different interests, different abilities and different goals—is it any wonder that we are often left confused? In planning for the introduction of [Australian Curriculum] science as a distinct subject [at Year 6 level], we had to look at what was expected of us across all areas of the curriculum. What extra content did we need to cover and how could we make this real for our students?

How to move from a dense to a mild fog

We know students bring their prior knowledge to a unit: 'Let's ask the students what they would like to learn.' We could then use this knowledge to create a unit of work . . . Linking what the children wanted to know with the curriculum expectations was the next challenge . . . They were asked to design an object (design and technology) that contained at least one working circuit (science content), create a budget and explore how their object could be sold in the real world (economic content) . . . [and] stage an expo . . . so that they could show and explain what they had learnt from a science perspective (oral language perspective) . . . How could we ensure that they had the basic skills and understanding to meet the project requirements? 'Let's ask the students what they needed to learn first.'

Implementation: From mild fog to gentle haze

While the students were grasping the scientific concepts, it was not that easy for us as teachers either. The question, 'Why is it not working?' was enough to bring back the dense fog and confusion. I had no ideas. We had connected the red to the positive, black to the negative, and no wires were touching, or so we thought—but nothing! 'Let's explore that together and see if we can find a solution' seemed to work. Within a short time, we had a group of students engaged in trying to solve the problem. We had 'experts' coming in to explain their thoughts regarding a solution, groups were trialling different options and at the end of the lesson we all learnt the importance of making sure that no two wires touched even slightly! It is okay not to know everything and to ask for help from experts, and yes, sometimes it is the students in your class who will be the experts. This is something that cannot be planned—you need to embrace your confusion and model a solution.

Evaluating: From gentle haze to clarity
A few things became clear. As a wise teacher once said, 'Confusion is a good thing.' It makes you reflect on what and why you are doing things. It helped to identify the need for integration across the curriculum. It helped to shape a lesson that allowed students to develop confidence in creating ideas, challenging content and finding solutions . . . we need to teach the curriculum but we can't forget that students need to be engaged and given the opportunity to explore, experiment, modify and learn through their own experiences.

Van Corler (2013), pp. 71–2.

The advanced organiser (Figure 4.1) asks five questions, related to intended, implemented or experienced curriculum. We will now address these in turn.

Question 1: What is possible?

Take a moment to consider what is likely to be considered when creating a science curriculum with the sources/possibilities/pressures that could be considered.

Sources/possibilities/pressures

At the top of Figure 4.1 are the sources of pressures and possibilities for the science curriculum. These include the following:

- *Societal expectations.* These are in line with society's expectation that an education will provide its children with the knowledge, skills and abilities to become active and successful adults. Stakeholders in this group include governments, business groups, universities, community groups, parents, students and teachers. Each stakeholder can bring different ideas and perspectives about what is needed or desired or expected within the science curriculum, and at times these can clash and other times reinforce others; sometimes stakeholders hold and express very passionate views. For instance, put yourself in the shoes of a parent. What could they want that is best for their child? How could this compare with a small business or a community group?

- *Knowledge organisation and development.* This includes what knowledge (inclusive of skills and attitudes as the dimension of competence is included) constitutes worthwhile learning, and in what forms and to what extent it will be used. Again, what is worth learning and why? Like all the different sources/possibilities/pressures, Layton's (1994) four pressures, Fensham's (1985) different demands and Reiss's (2007) aims should be considered again, demonstrating that political, economic and cultural pressures can impact what is considered worthwhile learning.
- *The nature of children and adolescents.* This entails some recognition of what they bring to learning situations and the diversity of those experiences; and ideas about learning, which was explored in more detail in Chapter 2.

What could be included in the curriculum?

Given all the sources, what could be included in any science curriculum is beyond the scope of what is possible. Hence some compromises need to be made about what is included. Not everything can be included, as you will note with the arrows blocked off in Figure 4.1.

Traditionally, science curricula around the world have remained focused on four strands of science: biology, chemistry, earth and/or environmental science and physics. This has included elements of scientific processes, the nature of science itself and the context in which science collaborates and works. There has been little inclusion of emerging areas of science, such as psychology and neuroscience—for example, imaging science—and even biochemistry, which has been in existence for approximately 100 years but is only now beginning to creep into science curricula.

Question 2: Why does the curriculum look like it does?

As James, a pre-service teacher, mentioned at the start of this chapter, curriculum can alter between states and territories and countries. Comparing different curriculum policy documents (*intended curriculum*) can help us to understand our own, including recognising what we do or don't value about the curriculum we are using.

The school curriculum: What's in the Australian Curriculum?

The Australian Curriculum is an example of an *intended curriculum*. It looks at some of the desired inclusions and pressures, and provides an overview of the structure and why it looks the way it does.

The Australian Curriculum was devised by a curriculum authority overseen by the federal government. However, in Australia the state/territory governments are responsible for their curricula. Except for Victoria and New South Wales, all states and territories have adopted the Australian Curriculum. Curriculum authorities in Victoria and New South Wales have made significant changes—and different changes in each state. This is a reflection of the different pressures within the states and territories across Australia, including different political viewpoints and maintaining traditional subjects. Furthermore, in Victoria, non-government schools enjoy a high degree of autonomy and can choose between implementing the Victorian or Australian Curriculum. Government and Catholic schools must implement the Victorian curriculum.

The rest of this chapter will focus on the Australian Curriculum, since this is the main curriculum taught in Australia. You will be able to recognise elements of this curriculum in the Victorian and New South Wales curriculum, but more interesting are the differences. While not discussed in this chapter, we recommend you take time to look over the state and Australian Curricula, especially if you plan to teach in Victoria or New South Wales.

In the Australian Curriculum, the clear intention has been to provide students with the opportunity to develop scientific literacy, 'helping them to understand more about science and its processes, recognise its place in our culture and society, and be able to use it in their daily lives' (Goodrum & Rennie, 2007, p. 3) and become effective citizens. This is based on the objective from the Australian School Science Education National Action Plan 2008–12 (Goodrum & Rennie, 2007, p. 14):

> to provide all Australian school students with access to science curricula that are relevant, contemporary and can be adapted to the needs of students in diverse situations to ensure that all students have opportunity to develop scientific literacy and become effective citizens.

Inclusivity is an important societal consideration for Australia, with its multicultural makeup as well as its recognition of Aboriginal and Torres Strait Islander cultures as part of the social fabric. The curriculum policy documents play a leading role in this. Celebrating diversity among people and providing opportunities for *all* students to learn is a fundamental goal of our democratic society. The National Action Plan drew on Western Australia's Curriculum Council report to outline what it meant by inclusivity:

> Inclusivity means providing all groups of students, irrespective of educational setting, with access to a wide and empowering range of knowledge, skills and values.

> It means recognizing and accommodating the different starting points, learning rates and previous experiences of individual students or groups of students. It means valuing and including the understanding and knowledge of all groups. It means providing opportunities for students to evaluate how concepts and constructions such as culture, disability, race, class and gender are shaped. (Curriculum Council of Western Australia, 1998, p. 17)

Essentially, curriculum takes the shape it does because it is considered to be relevant to students at that particular time and place. In considering relevance, Stuckey and colleagues (2013, p. 19) state 'science learning becomes relevant education whenever learning will have (positive) consequences for the student's life'. Positive consequences can include fulfilling students' current and future personal interests and educational demands. These positive consequences are also dependent on the values—those principles that guide a person's decision-making, beliefs and attitudes and are connected to personal integrity and identity (Halstead, 1996)—that students and their teachers bring to the classroom. Values influencing this notion of relevance will have both intrinsic components such as students' interests and motives, and extrinsic components such as 'ethically justified expectations of one's personal environment and the society in which they operate and live' (1996, p. 20).

Alignment of students' values with the values of science education is often not considered, even though the dimension of the nature of children and adolescents is an important source of social pressure on the science curriculum. It is usually adults—such as curriculum developers who assign values underpinning curriculum, as has been evidenced by the continued emphasis on traditional science—who make the decisions. Traditional science as represented in school curriculum relies on the out-dated belief that school science relies on a body of knowledge in the form of concepts that can be supported by data generated through experimentation. Such a common value held by many involved in school science has been extremely difficult to shift, partly because this is the very system that has rewarded teachers as successful science learners in their own schooling and university study.

Contemporary science, as opposed to traditional science, relates to more recent and emerging areas of science research and current issues. This was outlined in the *Shape of the Australian Curriculum: Science* report (ACARA, 2009, p. 5):

> Contemporary science involves new and emerging science research and issues of current relevance such as energy resources and technology, climate change and adaptation, mining and minerals, biodiversity and ecological sustainability, materials science and engineering, and health and prevention and treatment of disease.

Almost ten years later, there are many new and relevant issues. However, issues such as energy resources and technology, climate change and others are still valid today—although the notion of technology and its ability to be 'disruptive' may have changed the ways in which such issues are viewed. Again, the notions of relevance are highlighted here, reinforcing the important intrinsic and extrinsic sociological perspectives of contemporary science. Keep these aspects in mind when considering the structure of the Australian Curriculum: Science.

The structure of the Australian Curriculum: Science

The structure of the Australian Curriculum: Science includes eight learning areas, seven general capabilities and three cross-curriculum priorities spanning primary and secondary schools. The learning areas, general capabilities and cross-curriculum priorities are intended to be taught in an integrated way. Figure 4.2 shows the three-dimensional nature of the Australian Curriculum. This chapter is interested in the learning area of Science, integrated with the general capabilities and cross-curriculum priorities. For instance, Critical and Creative Thinking are significant aspects of science.

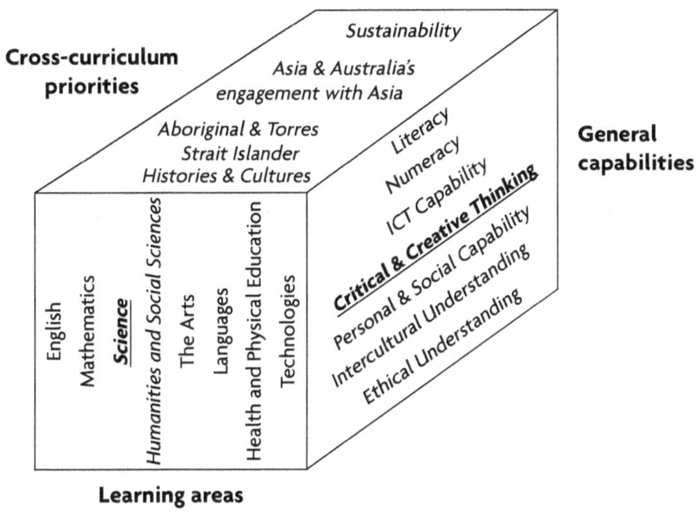

Figure 4.2 The three-dimensional Australian Curriculum
Source: Modified from Kidman & Casinader (2017).

The Science Curriculum is structured around three key inter-related elements:

- *science understanding (SU)*—the selection and integration of appropriate science knowledge (located in biological science, chemical science, earth and spaces sciences and physical science) in ways that explain and predict phenomena, and apply that knowledge to new situations
- *science inquiry skills (SIS)*—the skills associated with science inquiry, such as posing questions, planning, conducting and

critiquing science investigations, evaluating claims made with evidence and communicating findings
- *Science as a human endeavour (SHE)*—humans seek to improve their understanding and construct explanations of the natural world though the use of evidence and knowledge that is dynamic in nature. Science influences society by posing and responding to social and ethical questions, while science funding is influenced by the needs and priorities of society. Science is a unique way of knowing and doing, and plays an important role in contemporary decision-making and problem-solving. In making decisions about science, ethical and social implications must be taken into account. Science advances through the contributions of many different people from different cultures, and provides many rewarding career paths.

Within the structure of the Australian Curriculum: Science, there are different foci and some unifying ideas emphasised at different stages of schooling. These are indicated in Table 4.1.

The focus and unifying ideas provide some guidance around expectations about what is appropriate at different stages of schooling. That said, the unifying ideas in the earlier years remain important and useful in the older years. For instance, a focus on the science practices of observation and questioning is emphasised in K–2 but is just as relevant in 7–10 and beyond.

The Australian Curriculum: Science indicates the intended learning outcomes for the youth of Australia in terms of science and particularly scientific literacy, making up the *intended curriculum*. It is now time to turn to the *implemented (or taught) curriculum* that teachers create.

Questions 3 and 4: What happens in the school?

How does a teacher bring the curriculum to life for their students? In Figure 4.1, the *implemented curriculum* is represented by the

Table 4.1 Focus and unifying ideas at different stages of schooling

Stage of schooling (approximate age range in years)	Focus and unifying ideas
K–2 (4–7 years)	Focus on awareness of self and the local world with unifying ideas of exploration, observation, order, change, questioning and speculation
3–6 (8–11 years)	Focus on recognising questions that can be investigated scientifically and investigating them with unifying ideas in patterns, systems, relationships, evidence and explanations
7–10 (12–15 years)	Focus on explaining phenomena involving science and its applications with unifying ideas of sustainability, equilibrium and interdependence, form and function, evidence models, explanations and theories
11–12 (16–18 years)	Focus on disciplines of science—physics, chemistry, biology, earth and environmental science

planning for learning, and the teaching and learning processes. Teachers are responsible for developing and implementing a curriculum that suits the needs of their students to create meaningful experiences—a task that is often innovative, challenging and ever-changing, but at the same time immensely rewarding.

Planning for learning experiences; teaching and learning approaches

Case Study 4.1 represents the story of a Year 6 (primary) teacher who was part of a team developing a new integrated unit based on the Australian Curriculum. While, on face value, many secondary teachers would not find an integrated unit to be of much use as an example, these approaches to implementing the curriculum are becoming more prevalent in all stages of schooling. The fact that the Australian Curriculum is three-dimensional, as pointed out in Figure 4.2, points to the need to plan integration of learning areas, cross-curricular priorities and general capabilities.

As discussed earlier, the Australian Curriculum: Science consists of many different components. The learning areas, general capabilities and cross-curriculum priorities are intended to be taught in an integrated way, as are the three interrelated science strands (SU, SHE and SIS). Teachers need to consider ways to teach science in an integrated way, and avoid teaching the curriculum as isolated, unrelated and separate segments. The unifying ideas may offer a way to do this. For instance, the unifying ideas include a number of science practices, such as observation, patterns, explanations, models and systems. A focus on teaching science concepts *with* science practices offers a way for teachers to connect the curriculum and build on students' learning experiences within and across years (Marangio, 2013). Teaching the science concepts with the science practices that developed these concepts can highlight why and how science develops its knowledge, and therefore embrace SU, SHE and SIS and connect to the general capabilities.

Question 5: What is the role of assessment?

The final curriculum aspect to be considered from Figure 4.1 is the *experienced (realised) curriculum*, or the outcomes of the implemented curriculum. An integral part of this is what is assessed, and this element is explored in more depth in Chapter 7. Equally important is the role of evaluation; however, mechanisms are needed whereby teachers evaluate the success or otherwise of how their implementation of the intended curriculum has been realised by student learning outcomes.

Intentional: What is assessed

The expo event that was part of Case Study 4.1 provides a tangible form of evaluation for teachers. In addition, the reflective elements also provide some evaluative insights about the impact of this learning experience. Articulating it for others, such as in Case Study 4.1 (Van Corler, 2013), also provides a more critical lens of such reflection. Questions such as 'How will I know if the learning intentions of the curriculum have been realised?' are important evaluative questions as the teaching and learning processes are developed. This means not only that the planning of the experiences needs to occur, but planning the assessment and evaluation aspects also needs to happen concurrently at this time.

Unintentional: The hidden and null curriculum

Equally important in examining this case is to see what other unintended consequences of this learning experience are created. This is part of the experienced curriculum, which can be broken up into two other types of curriculum that we need to explore. The first is the **null curriculum**—those things that are 'systematically excluded, neglected or not considered' (Joseph et al., 2000, p. 4). In many instances, the student voice in making curriculum decisions is often part of the null curriculum. This is not so in Case Study 4.1, and the inclusion of student voice or expertise was fundamental to the success of this learning experience. The second type of curriculum we need to explore is the **hidden curriculum**, or the 'outcomes from teaching-learning activities that are not part of the explicit learning intentions of those responsible for the planning of those activities' (Smith & Lovat, 2003, p. 34). In Case Study 4.1, the hidden curriculum included the realisation that confusion can be a positive, and that expertise can reside in unexpected places.

null curriculum part of the experienced curriculum; relates to what is absent, overlooked or left out, and therefore can send powerful messages to students (whether intended or not)

hidden curriculum part of the experienced curriculum: the informal and often unintended learning experience, such as hidden messages, values, attitudes and beliefs conveyed in the classroom

When considering the *experienced curriculum*, think about the lasting learning that your students may realise—not just measured in assessment, but also the null and hidden learning that may be present in the weeks, months and years that follow. What do you consider is worthwhile learning? What impressions do you wish to leave with your students long after they have been taught by you?

CONCLUDING COMMENTS

Earlier in this chapter, the notion of alignment of students' values with the values of science education was often not considered as part of the *implemented curriculum.* While this is implicit in Case Study 4.1, some examples (one primary and one secondary) of where such values are explicitly aligned can be found in Corrigan and Smith (2015). In these examples, the focus for the *implemented curriculum* is on what is common about values such as creativity, open-mindedness, logical thinking in science and in the ways students engage with different knowledge domains of the curriculum—particularly science as a human endeavour and science inquiry skills.

The term 'curriculum' can mean many things, and relate to the entire school educative program within a classroom, school, state/territory or country, as well as outside school. This chapter has used the Australian Curriculum: Science as an example of *intended curriculum* to discuss curriculum development, implementation and experience. While other curriculum is used in Australia and elsewhere, the principles remain the same. As a teacher, you play a key role in creating powerful and relevant learning experiences for your students. What you do as a teacher matters and can have a lasting impression on your students. You need to consider how these elements of the curriculum are framed into positive experiences for our children to assist them to become active and successful adults.

DISCUSSION QUESTIONS

4.1 Does the Australian Curriculum: Science reflect society or does it attempt to initiate change in society?

4.2 What evidence can you see that a theory of learning is evident in the Australian Curriculum: Science?

4.3 Who are the shapers of curriculum?

4.4 What criteria would you use to assess quality teaching in this curriculum?

4.5 Consider the 'hidden curriculum'. What are some of the factors that make up the hidden curriculum?

FURTHER READING

Aubusson, P.J. (2011). 'An Australian science curriculum: Competition, advances and retreats'. *Australian Journal of Education*, 55(3), 229–44.

Corrigan, D. (2018). 'The Australian science curriculum: Realizing the opportunity?' In A. Reid & D. Price (eds), *The Australian Curriculum: Promises, Problems and Possibilities*. Canberra: ACSA, pp. 68–85.

Corrigan, D., Dillon, J. & Gunstone, R. (eds) (2007). *The Re-emergence of Values in Science Education*. Rotterdam: Sense.

REFERENCES

Apple, M.W. (1979). *Ideology and the Curriculum*. London: Routledge and Kegan Paul.

Australian Curriculum, Assessment and Reporting Authority (ACARA) (2009). *Shape of the Australian Curriculum: Science*. Canberra: ACARA.

Bobbitt, J.F. (1918). *The Curriculum*. Retrieved from https://archive.org/details/curriculum02bobbgoog.

Corrigan, D. & Smith, K. (2015). 'The role of values in teaching and learning science'. In J. Deppler, T. Loreman, R. Smith & L. Florian (eds), *Inclusive Pedagogy Across the Curriculum* (pp. 99–118). Bingley: Emerald Books.

Curriculum Council of Western Australia (1998). *Curriculum Framework*. Perth: Curriculum Council of Western Australia.

Fensham, P.J. (1985). 'Science for all: A reflective essay'. *Journal of Curriculum Studies*, 17, 415–35.

Gintis, H. (1972). 'Towards a political economy of education'. *Harvard Educational Review*, 42(2), 70–96.

Goodrum, D. & Rennie., L. (2007). *Australian School Science Education National Action Plan, 2008-2012 Volume 1: The National Action Plan.* Canberra: Commonwealth of Australia. Retrieved from http://apo.org.au/system/files/4048/apo-nid4048-45771.pdf.

Halstead, J.H. (1996). 'Values and values education in schools'. In J.M. Halstead & M.J. Taylor (eds), *Values in Education and Education in Values*. London: Falmer Press.

Joseph, P.B., Bravmann, S.L., Windschitl, M.A., Mikel, E.R. & Green, N.S. (2000). *Cultures of Curriculum*. Mahwah, NJ: Lawrence Erlbaum.

Kidman, G. & Casinader, N. (2017). *Inquiry-based Teaching and Learning Across Disciplines: Comparative Theory and Practice in Schools*. New York: Palgrave Macmillan.

Layton, D. (1994). 'STS in the school curriculum: A movement overtaken by history?' In J. Soloman & G. Aikenhead (eds), *STS Education: International Perspectives on Reform*. New York: Teachers College Press, pp. 32–44.

Marangio, K. (2013). 'Progression of learning of psychology within the Victorian Curriculum', unpublished MEd thesis, Monash University.

March, C. & Willis, G. (1995). *Curriculum: Alternative Approaches, Ongoing Issues*. Englewood Cliffs, NJ: Merill.

Pinar, W. & Grumet, M. (1976). *Toward a Poor Curriculum*. Dubuque, IO: Kendall/Hunt. [Reissued in 2000 as *Curriculum Theorizing: The Reconceptualization*, by Educator's International Press.]

Posner, G. (1995). *Analysing the Curriculum*. New York: McGraw-Hill.

Reiss, M. (2007). 'Aims of school science education'. In D. Corrigan, J. Dillon & R. Gunstone (eds), *The Re-emergence of Values in Science Education*. Rotterdam: Sense, pp. 13–28.

Smith, B., Stanley, W. & Shores, J. (1950). *Fundamentals of Curriculum Development*. New York: Harcourt Brace.

Smith, D.L. & Lovat, T.J. (2003). *Curriculum: Action on Reflection*, 4th ed. Sydney: Social Science Press.

Stuckey, M., Hofstein, A., Mamlok-Naaman, R. & Eilks, I. (2013). 'The meaning of "relevance" in science education and its implications for the science curriculum'. *Studies in Science Education*, 49(1), pp. 1–34.

Van Corler, L. (2013), 'Confusion is a good thing: Even in teachers' in S. Keast, G. Lancaster, J. Loughran & D. Panizzon (eds), *Teaching and Learning Science Through Cases*. Melbourne: Monash Print Services, pp. 71–2.

Waring, M. (1979). *Social Pressures on Curriculum Innovations: A Study of the Nuffield Foundation Science Teaching Project*. London: Methuen.

Young, M.F.D. (1971). 'An approach to study curricula as a socially organized knowledge'. In M.F.D. Young (ed.), *Knowledge and Control*. London: Collier Macmillan, p. 19–46.

CHAPTER 5

THE NATURE OF SCIENCE

KATHY SMITH, DEBORAH CORRIGAN *and* **JENNIFER MANSFIELD**

LEARNING OUTCOMES

After reading this chapter, you should be able to:

- articulate what the nature of science (NoS) is
- detail the implications of NoS for science learning and teaching
- consider the possibilities and challenges with which primary and secondary educators might grapple when incorporating components of NoS into their own practice.

Nature of Science (NoS) is a phrase that you may have come across at different times as you engaged with literature and research related to science learning and teaching, particularly in terms of what might constitute quality science learning and teaching. NoS is a construct often used in science education, as it provides us, as teachers, with a lens for making sense of what science is and the ways scientists work.

Indeed, NoS is an explanation of why science is structured in the way it is and is different from other disciplines—for example, history—each of which has its own unique structures. But before unpacking NoS further, consider the case study below. This case provides first-hand insights into one way in which scientists go about generating new knowledge.

CASE STUDY 5.1 How do you get mosquito spit and why would you want to?

Scenario: Setting the scene

A research team of scientists at Monash University, Melbourne, led by Professor Beth McGraw, wants to know more about the transmission of dengue fever. (This story can be accessed at http://www.monash.edu/education/events/stem-institute/stem-talks.html.)

Reason: Why invest time and effort?
Approximately 380 million people in developing countries around the world are at risk of contracting dengue fever, a potentially deadly virus that has no vaccination or cure. Beth's team wants to understand how the virus is transmitted so the spread of the disease can be managed.

Research questions: Most scientific endeavours start with one and they need to be investigated using a scientific approach
How long does it take for a mosquito that has just bitten a person with dengue fever to spread this virus to another human? For how long might the same mosquito keep spreading the disease to each new person it bites?

Drawing on existing knowledge to help guide the design of the scientific investigation to test our prediction
We think it might take about 10 days before the virus is able to be transmitted by a mosquito and a mosquito might continue to secrete the dengue virus throughout the rest of its life.

Research approach: A series of steps
Step 1
Infect a large number of mosquitoes with dengue fever. Test when they start to secrete dengue fever in their spit and for how long they might secrete the virus. This sounds easier than it probably is! There were four problems that the team needed to solve before work could begin (see Table 5.1).

Step 2
Feed virus cultured blood meal to the hundreds of female mosquitoes. Collect caps from individual mosquitoes and test for the presence of dengue fever in the spit. Continue experiment till mosquitoes die. Remember—work carefully and keep the doors closed so you don't get bitten!

Stop the press—disaster strikes!!!

Dear reader, unfortunately, after months of preparation, on the day the team was going to start the experiment, the researchers found that the virus culture was contaminated with bacteria! Eight months of preparation and creating designed solutions to the problems were thwarted by a contamination issue that had not been foreseen. All the mosquitoes died in vain and the virus culture was discarded. Months of work went down the drain.

Step 3
Pick ourselves up, dust ourselves off and start the whole process again, but this time being more aware of possible contamination issues.

Success! The second time around, all the experiments went according to plan. Due to the unforeseen setback, the team worked faster and better the second time around. There was value in learning from failure.

Table 5.1 Problems that needed solving before work could begin

	Problem 1	Problem 2	Problem 3	Problem 4
Question that requires a solution	How to obtain mosquito spit without killing the mosquito	How and where to house hundreds of dengue-infected mosquitoes for a couple of weeks	How to breed hundreds of mosquitoes	How to create large quantities of dengue-infected blood meal to feed the mosquitoes
Designed solution	Months of trial and error gave rise to small cups (they look like small soft drink bottle lids) that they could remove every couple of days and collect the hundreds of individual mosquitoes' spit.	Months of trial and error gave rise to little 'apartments' (i.e. a small beaker covered with gauze). The cups with spit could be removed and replaced easily.	It takes a month to grow a mosquito from egg to adult. Within three months of breeding, the team could have three generations of mosquito females ready to be infected with dengue fever (only females eat blood meals).	Grow large quantities (litres) of cultured mosquito cells in the lab to culture the virus.

The findings: Uncovering the unexpected

What we found was that, unlike the prediction of it taking 10 days for the virus to be present in the saliva, it only took 4 days! That really helped explain how mosquito borne outbreaks can travel so quickly. We also found out that the virus was only present in the saliva for 10 days, not the mosquito's whole lifetime. Indeed, it was the middle age females who were the transmitters of the disease.

The team had posed informed research questions, planned experiments requiring creative designed solutions which arose through trial, error and perseverance, ran into difficulty, learnt from setbacks, gathered data, posed explanations and shared their findings with the greater community.

Beth showcased her team's research during the Monash University *STEMTalks®* series during Science Week in 2015. You can access the video via the following link for further insights into this work: http://www.monash.edu/education/events/stem-talks/#videos.

Case Study 5.1 highlights a modern success story for scientists in their fight against dengue fever. This story is peppered with success and failure for the research team. For example, after eight months of setting up, seeing all their work amount to nothing must have been devastating. We often forget the people who conduct this work: why they feel compelled to do the work, how they come to know what is important and how they decide what to do with that knowledge. These important aspects underpinning scientific work, such as posing informed research questions, planning investigations, creating designed solutions to problems, working through trial, error and perseverance, gathering data, posing explanations and communicating findings to a wide range of audiences, are what we mean when we refer to NoS.

In this chapter, we will explore the important theoretical and practical aspects of NoS through two key questions:

1 What is the nature of science (NoS)?
2 What are the implications of NoS for science learning and teaching?

These two key questions will frame considerations about what we, as individuals, understand science to mean and whether this is similar to the ways in which others perceive science. When we view the world through a science lens, we are applying a particular way of knowing to situations. If there are differing views about what science is, how is it that we can come up with some consistent explanations for what we observe around us? This is a particularly pertinent question to consider as teachers of science, across both primary and secondary schooling, because it prompts deep thinking about the impact of NoS on ways of working and the conditions fostered in classrooms—in terms of both possibilities and challenges.

We will begin our exploration of the complexities and authenticity inherent in NoS and its role in science education by considering the first key question: What is NoS?

KEY QUESTION 1: WHAT IS THE NATURE OF SCIENCE (NoS)?

Before launching into a more direct response, we invite you to think back to Case Study 5.1. Using this work as a reference point, pause and consider what science is to you and what are some of the features that might characterise science practices and scientific knowledge. Without detailing your insights just yet, this train of thought invites a consideration of the issues that underpin what is widely recognised in historical, philosophical and educational discourse as the *nature of science* (NoS). In science education literature, a frequent response to such pondering simply states that science is a way of knowing the world (Lederman, 1996). Yet understanding and conceptualising this idea is not without complication because the notion of 'knowing' fundamentally concerns a way of thinking that combines a complex mix of values and beliefs inherent to an individual and the development of their own scientific knowledge. Science can also embody a way of acting because values and beliefs provide the guiding principles, behaviours and more immediate responses that underpin how knowledge is developed. This section will now explore this notion of NoS and various ways in which it is interpreted and understood by the science community in particular.

NoS out in the world

Everyday life exposes the public to wide-ranging representations of science, whether this is through school science, the media, technology or the public face of science—largely through science professionals such as doctors, nurses, technicians, academics, mechanics and engineers. Society is constantly interacting with science in many forms. The capacity to recognise the characteristics that become the common thread of 'science' in all these situations may be challenging, because scientific enterprise as it is encountered in everyday life is embedded, complex, diverse and yet simultaneously very specialised. As a result, science can and does mean different things to different people. Yet for those who work as part of the scientific community, science frames not so much the diversity or the complexity of the work they undertake but the fundamental tenets and principles that characterise all scientific work, regardless of field or discipline. For those involved in scientific endeavour, science is a way of knowing, a way of thinking about and acting in the world around them. It fundamentally involves the importance of observation, questioning, collecting and analysing data (generally **empirical data**), making **inferences**, providing justification, generating explanations (often in the form of models and systems), communication

empirical data information that has been directly or indirectly observed

inference reaching a conclusion or explanation based on the data and evidence available

and **peer review**, where experts in the field critically review the work of their colleagues in the same field. For those in the scientific community, science refers to the very nature of their human endeavour—that is, the processes involved in the social construction of knowledge, as captured in the opening case study.

> The cultures in the various areas of experimental science differ greatly, though we all use a language based on **hypothesis**, experiment, observation, **open reporting** and **independent verification**. Even if the world of cosmic scientists probing the beginning of the universe seems very different from that of the research biologist, the common link is that we all come down to dealing with new data. (Doherty, 2005, p. 60)

In this quote, Peter Doherty, a Nobel Prize winner for research involving infection and immunity, is talking about **experimental science**, which again is a particular type of work conducted in science. Other types of science include more **descriptive investigations**, such as **fieldwork**, or are correlational in nature as they attempt to establish relationships between two or more things.

peer review the process by which research is critiqued critically by scientists in the same field of research

hypothesis a proposition or proposed explanation made on the basis of evidence that is currently available as a starting point to investigate further

open reporting an important part of peer review where scientists present their data in scientific journals where others scientists can critically review their work

independent verification an important underlying principle of peer review where experts in your field independently critique your work (in a process that is often blind in that the reviewers do not always know the author of the work they are reading), verify (or not) its accuracy and agree (or not) with the argument presented

experimental science a science practice focused solely on conducting experiments

descriptive investigation scientific work focused on describing what is observed

fieldwork science investigation that take place out in the field

In terms of our understandings of NoS as science educators, this means that we need to recognise that it can be diverse and dynamic because of the human elements involved. However, there are some central tenets involved, in that NoS provides us with a particular lens for knowing, thinking about and acting in ways that will help us to make sense of our world.

Teasing out NoS

Science is an intellectual pursuit based on a premise that the things and events in the universe occur in consistent patterns that are comprehensible through careful, systematic study. Science therefore embodies the pursuit of this knowledge production, and as part of a science community scientists seek to undertake this work, sharing certain basic beliefs and attitudes about what they do and how they view their work.

Given the complexity of historical, philosophical and educational perspectives, a singular and agreed definition of the nature of this activity has not been reached. Yet enough agreed generalities exist to define and characterise science—or at least a general understanding of science—as a human endeavour (Osborne et al., 2003; Schwartz, Lederman & Crawford, 2003). For example, scientific knowledge is accepted as being both simultaneously reliable and tentative. This notion means it is possible to have confidence in existing scientific knowledge while also acknowledging that such knowledge is subject to change with new observations and as existing observations are reinterpreted and reconsidered. In Case Study 5.1, this plays out through the well-considered and implemented research design—for example, how to keep the mosquitoes alive—but is also evident in the unknown nature of the final outcome (the virus took only four days to turn up in the saliva rather than ten days, and only lived for ten days). The acceptance among scientists of tentativeness is part of a self-correcting aspect of science, which in itself

theory a set of ideas that intends to explain some phenomenon; a theory cannot be proven, but rather is disproven, as it is a set of ideas that are created in order to explain

law a description of an observed phenomenon; a law does not intend to explain (as a theory does), but rather to describe the phenomenon

becomes reliant on scientific knowledge that is based on and derived from observations of the natural world. This is where we start to see human endeavour in action. The work of scientists is influenced and driven by currently accepted scientific **theories** and **laws**.

These two terms indicate similar but different kinds of science knowledge. Scientists refer to current theory and use laws to develop questions, investigations and interpretations of data, which in turn help them to shape and determine the work of science and its progress as well as to maintain consistent knowledge development. Science is also based on both observation and inference, which are subjective in their nature due to the human elements inherent in who is observing and making the inferences. For example, when a group of people watch two identically sized balls of different weights drop from a height of 10 metres, they will tend to see what they expect. Some will see the balls land together, while others will see the heavier ball land first.

In the relationship between observation and inference, creativity becomes an important component of scientific knowledge as imagination and logical reasoning underpin observations and inferences of the natural world. In the ball example given above, those people with a more limited understanding of forces or less experience of forces such as gravity may expect the two balls to land at the same time. However, those with an understanding or experience of forces such as gravity could reason that the heavier ball would have more mass on which gravity could pull, and therefore would land before the lighter ball. Therefore, subjectivity influences working with existing current theory through the personal values, agendas and prior experiences of those conducting the work. Science is a human endeavour, and as members of society, scientists are influenced by the culture in which they practise. While the example of the balls falling to earth above is a simple one, our understandings are intimately linked with our experiences, whether they are scientific or cultural.

To revisit some of the points made in response to this first key question, scientific activity is culturally embedded, reliant on empirical observation and subject to change, with each of these activities working together interdependently (Schwartz, Lederman & Crawford, 2003). Each of these acts is influenced by the culture and society in which the science is practised, as well as by the frameworks and personal subjectivity of the scientist.

Communicating what NoS is and why it matters

Certain practices and behaviours underpin the framework within which scientists operate. One example of this is the importance of communication and peer review among scientists as an effective form of accountability and self-regulation. Case Study 5.1, which originally was presented as a 15-minute TED Talk-styled presentation, is an example of making science, including NoS and knowledge, accessible to the wider public.

The crucial act of certifying the results occurs through publication in scientific journals. The process of rendering results public through writing, and therefore in traceable, unalterable form, remains an important form of communication. Verification and certification of results and of the methods used constitute a process that is performed both as an essential service to other knowledgeable members of the scientific community and to the public at large. The peers in their own field of work depend upon the results of their colleagues. They acknowledge their indebtedness by citing the names of members of the scientific community, through competent mutual criticism and peer review. Scientists assume both responsibility for the reliability of their work via one another and public responsibility towards the wider society (Nowotny, 2005).

While it may be stated that scientists, as part of the scientific community, seek to share certain basic beliefs and attitudes about what they do and how they view their work, it cannot be so confidently asserted that all scientists think and behave in the same way. Science and society can never really be separated: women and men of various social and ethnic backgrounds engage in the activities of science, with some working in teams and some alone, but all communicate extensively with others. Scientists are influenced by societal, cultural and personal beliefs and ways of viewing the world (National Research Council, 1996). This notion of science as a human endeavour has not been the perception that has pervaded societal understandings of the work of scientists. While science (and by association scientists) is seen as objective, it can never really be objective, since it is humans working as scientists who provide interpretations and explanations for their observations. At best, science can be seen to reduce bias, where scientists recognise the bias they may possess and seek to reduce it as much as possible. Again, the process of publishing results in scientific journals that have been reviewed by peers in their own field of work helps to reduce such bias. A recent example has been the work of Professor Jennie Brand-Millar—the Australian scientist who battled years of criticism and scepticism in her work that developed a way to measure the body's absorption of carbohydrates—now known as the Glycaemic Index.

In principle at least, scientific findings that have been published should be able to be repeated by other scientists. In practice, this is rarely the case—even among specialists—due to the role played by tacit or unstated knowledge and other factors. The written form, which lends itself well to certification, and privileges a standardised form of communication (such as the standard template used for science articles in scientific journals or science reports in school), does not sufficiently capture local, practical knowledge and idiosyncrasies. Knowledge claims are considered temporary, since future findings may at any time overthrow accepted views (Nowotny, 2005).

These ideas about communicating NoS provide food for thought for us as science educators on a couple of levels. First, they might cause us to think about how we access contemporary information about the work of scientists, along with the ways they work, to share with our students in order to shed light on what NOS is in action. Second, they raise questions about how we might support students to document and communicate their own burgeoning understandings of NoS and what it looks like in practice through their own science learning.

Considering NoS through values

Rather than being objective and value-free, as depicted in some school science texts and settings, science has a range of values that underpin it as a unique way of knowing and acting, which is what makes it recognisable as a discipline in its own right. Table 5.2 presents some of the values that underpin science as it focuses on only some of the **epistemic** (the knowledge of how knowledge is) and sociological (all those factors that relate to society) dimensions of values of science.

However, providing definitions of fundamental values that underpin science contributes to understandings of NoS. These values are often not explicitly considered in science education and provide an opportunity to appreciate the attributes of science as a powerful way of knowing and acting. In Table 5.2, the heading 'constructed meaning' highlights the fact that there is no

epistemic knowing how something is known. Epistemic values in science include scepticism (the general attitude of doubting or questioning knowledge, opinions, beliefs or claims) or parsimony (when there are competing explanations, models or systems, the simplest explanation, models or system that accounts for most of the data are used)

one definition or meaning for these values, but rather a constructed meaning that represents some of the more acceptable understandings of these terms.

Table 5.2 Some values of science and their meaning

Value	Constructed meaning
Curiosity (includes intellectual curiosity and inquisitiveness)	Curiosity relates to not only the questioning but also to consideration of the wonder and mystery aroused when attempting to provide explanations for phenomena that occur in the natural world.
Parsimony	When there are competing explanations (hypotheses, models, systems, etc.), the simplest explanation, model or system that accounts for most of the data is accepted for use. Note: The term 'parsimony' derives from the Latin lex parsimoniae, which broadly translates as the law of parsimony, law of economy or law of succinctness. In more recent times, the phrase 'Ockam's razor' has been used to describe this value of science. The razor is a principle that suggests we should tend towards the simplest explanation until such simplicity of explanation can be substituted for greater explanatory power. The razor's value lies in the justification for deciding between the competing explanations. Such justifications also need to take account of the plausibility, fruitfulness and robustness of such explanations.
Creativity	Creativity—particularly in thinking—is characterised by its apparent lack of connection to other ideas. Processes such as 'random thoughts', 'thinking outside the box', 'guesses' and 'intuition' are often attributes of creativity. In attempts to promote the creative process, it is important not to place parameters around thinking.
Open-mindedness	Open-mindedness means being receptive to the alternative and different ideas and opinions presented by others. It is also attributable to looking at data in alternative ways.

Value	Constructed meaning
Rational thinking	Rational thinking, rationality or rationalism describe the cognitive process of reason: how one thought links to another. Rationality is different from logic in that it focuses on the notion of reason as a type of thought. Logic, on the other hand, is a process that attempts to describe the norms and rules by which reasoning operates, and in this way can attempt to teach orderly thinking. Logic is therefore one type of rational thinking. For example, rational thinking would include such things as skipping steps, working backwards or drawing diagrams—which are not part of logic.
Empiricism	Empiricism relates to knowledge that is derived from sensory experience (and includes direct and indirect observation). In this sense, it provides an alternative view to rationalism regarding the processes of knowledge creation, as rationalism is based on reason or thought. The power of these two seemingly opposing theories on knowledge creation lies in recognising their differences and making judgements based on the differences.
Scepticism	Scepticism relates to a general attitude of doubt or questioning of knowledge, opinions, beliefs presented as facts or claims that are made. Scepticism in science is often seen in the practice of subjecting beliefs and claims to scientific investigation (and further scientific investigation) in order to support their reliability.

Source: Corrigan (2015).

This set of values offers a frame, or a way of thinking about science, for educators in both primary and secondary settings to consider in terms of what they are fostering in their own students through their approaches to learning and teaching. It is worth considering how these values, and therefore how we articulate NoS, might be introduced and nurtured through our own classroom practice.

Summing up

A response to the question 'What is NoS?' can be represented in the following summary:

- NoS represents the discipline of science as a particular way of knowing, thinking and acting.
- Scientists, as part of a science community, seek to undertake their work sharing certain basic beliefs and attitudes about what they do and how they view their work.
- Scientists are people, and their thinking and activity are conducted within social contexts, which has a profound effect on the way science is done.
- Understanding NoS requires attention to the complexity of science as a human endeavour, which can involve a range of values and beliefs.
- Scientists' values and beliefs are influenced by the larger culture in which they live. Such personal views can, in turn, influence the questions they choose to pursue and how they investigate those questions.
- Scientists expect their colleagues and other scientists to scrutinise their ideas, evidence, methods and reasoning.
- The diversity of the scientific community helps to facilitate specialisation in different areas of science and provides different points of view that invigorate problem-solving and balance biases.
- Scientists expect that they and their colleagues will adhere to the ethical guidelines laid out by the community of scientists in which they operate.

This section has unpacked NoS and what it is from the perspective of scientists and the science community. This is important if we, as science educators, are to support our students to develop contemporary understandings of science and how it is practised. The next section examines how these ideas might be applied to the classroom by considering the second key question for this chapter: what are the implications of NoS for science learning and teaching?

KEY QUESTION 2: WHAT ARE THE IMPLICATIONS OF NoS FOR SCIENCE LEARNING AND TEACHING?

Case Study 5.1 illustrated one way in which scientists might practise science and the previous section teased out the components underpinning the nature of science, but this may raise questions about what this means in terms of classroom practice and how it is relevant to science learning and teaching. In considering the implications of NoS for science learning and teaching in both primary and secondary settings, this section first differentiates between science and school science before offering insights into how NoS might play out in practice.

Science and school science: Mind the gap

While the dengue fever research shared above highlights the complexities inherent in contemporary understandings of science practice, school science tends to simplify 'ways of knowing' so they are about students acquiring what scientists know (Duschl, 2000). With knowledge acquisition in mind, it is not uncommon for teachers to select activities and tasks that distil scientific facts and ideas into chunks of information. You may be familiar with this from your own experiences of learning science in school. In taking this approach, science learning is effectively disconnected from the context of how scientists have arrived at this thinking and diminishes insights into why they see one particular belief or conceptual scheme to be more effective or applicable than another. By doing this, school science under-estimates the complexity and implications of the embedded, or invisible, aspects of knowing that are essential to NoS. This limited curriculum emphasis eliminates any chance of students understanding the social, cognitive and epistemic dynamics that make science the way of knowing that it is:

> Our views about science itself have undergone substantial revision. Recent scholarship in the history, philosophy and sociology of science has effected a major shift from the view that scientific knowledge is universal, coherent, objective and unproblematic towards recognition that it is sometimes uncertain, contentious and unable to provide clear, unambiguous answers to many important questions. There is increasing recognition among science educators that science is a product of its time and place, inextricably linked with its sociocultural and institutional location, and profoundly influenced by its methods of generation and validation. (Hodson, 2003, p. 647)

Traditionally, the purpose and approach to science curriculum and the teaching of science in schools have valued academic outcomes (Fensham, 1988). For students, school science has often perpetuated a view of science as a rigid body of absolute unchanging truths consisting of isolated 'facts', devoid of human imaginations and logical reasoning, disconnected from society and culture and having no real application to everyday life. This is in contradiction, as shown in the earlier section, to how NoS is understood more broadly. In general, school science has reinforced inadequate educational representations of the complexities of science as a social activity, current practices in science and constructivist perspectives (as discussed in Chapter 2). With this limited focus, it could easily be argued that content knowledge can essentially become an abstract system of scientific knowledge (Fensham,

1988), lacking meaningful context and the relevance of real situations.

Pointing out this apparent gap between science and school science, in terms of understandings and practices, may seem like taking a deficit view of science education. Instead, we urge you to see it as a way to raise awareness and highlight where we need to focus our energies as science educators, both in primary and secondary schools. By acknowledging this disconnect, we are provided with an opportunity to reimagine what science education in schools could look like—particularly with regard to NoS and the place it has in our approaches to science learning and teaching.

Negotiating the gap

Understanding NoS can only be achieved when approaches to science learning and teaching acknowledge the importance of explicitly attending to the values that underpin the discipline of science. This can only be done by providing opportunities for students to think specifically about the nature of science and emphasising how various aspects—such as observation and inference—apply to the construction of scientific concepts, along with the implications this has for scientific knowledge in general (Bell, 2001). This type of science teaching attends to NoS as something to be 'planned for instead of being anticipated as a side effect or secondary product' (Akindehin, 1988, p. 73). Only then can the structure of science and the sociology of science be explored:

> Effective teachers do not leave to chance their students' development of accurate understandings of science content. Rather, they purposively plan and execute instruction designed to facilitate students' construction of scientifically valid understandings. Lessons on the nature of science should be no exception. (Bell, 2001, n.p.)

Yet this type of reflection or interrogation on the scientific enterprise is not typically encouraged by textbook-based science curriculum experiences (Bell, 2001). Teachers themselves need to develop a different way of thinking about and working with science as part of the school curriculum. They need to consider, understand, value and attend to the complex interdependency of science and people, demonstrating in their teaching practice that the ideas and conjectures they hear, know and explore are the result of people engaging in processes and activities of science—people observing, classifying, questioning, investigating and analysing. People construct the ideas of science. It is people who generate data and construct evidence to challenge and sometimes change existing ideas and understandings.

Most importantly, seeing science as a human endeavour allows teachers to potentially broaden their pedagogy to embrace contemporary approaches to science learning and teaching. There is increasing recognition of the need to look at the wider social, political, economic and ethical issues that surround the practice of science (Hodson, 2003). This type of science learning relies on opportunities to explore the value of negotiation and argumentation, collaborative thinking and communication, meaningful contexts for learning, opportunities for students to be engaged while exploring the diversity of opinion among not only themselves but the general public and the scientific community. Such experiences are not only important for quality science education but begin to reframe the ownership and use of science knowledge.

Implications of reducing the gap for classroom practices

Meaningful learning about NoS, while embedded in curriculum requirements, must essentially attend to student interests and learning needs. This requires alternative approaches to planning with regular reviews of the ways in which teaching is accessing, diagnosing and effectively addressing student thinking and learning needs. Flexibility becomes an essential aspect of teaching practice, enabling teachers to follow up particular student interests as they arise while still maintaining focus on the overall intended learning outcomes. However, bringing this level of awareness to science learning and teaching on its own is not enough. Intertwined with this student-centred focus should be attention to NoS in ways that are aligned with contemporary approaches to science and that require attention to be given to a number of key aspects of learning and teaching. While many considerations may be at play, three areas will be discussed briefly in terms of how they might impact classroom practices without being prescriptive.

Building on knowledge, skills and understandings

The area of knowledge, skills and understandings requires students to develop a range of thinking and learning behaviours that reflect how scientists really work. For example, if we return to the earlier example of dengue fever research (Case Study 5.1), Beth and her team were required to draw on a range of knowledge, evidence and persistence to solve the problems with which they were faced. In a classroom setting, achieving this means fostering learning experiences that explicitly emphasise the value of reflecting and planning, thinking and working creatively, communicating and collaborating, exploring, experimenting and creating new knowledge. As Hildebrand (2007, p. 46) states, 'the science curriculum is the bequeathing of a set of cultural practices from one generation to another', and this in turn means that science is not so much about developing a set of recipes and procedures, which tends to be a common experience—especially in secondary school settings. This is not such an issue in primary schools, where learning skills are usually fostered effectively, but the greater challenge is engaging with conceptual science knowledge to inform these learning behaviours. Science learning experiences must provide authentic representations of the skills and behaviours that are valued and practised as an embedded part of 'knowing' in science. This, however, requires a shift in present teaching emphasis:

> Many students believe that science is all about dramatic discovery, rather than generating knowledge and creating explanations through repetitive work and the painstaking collection of evidence. When conducting science investigations students rarely think about the 'thinking behind the doing' and the need for evidence that will be believable and acceptable to others. (DEECD, 2006)

One resource upon which primary and secondary teachers can draw to assist with this more integrative approach to science learning and teaching is the *Science Continuum* (DEECD, 2006). This online science resource emphasises the need for science teaching that has an explicit pedagogical purpose and that, by design, attempts to develop new emphases for science learning at all levels. Overall, these strategies are designed to enable students to take ownership of scientific knowledge and make sense of them in ways that are meaningful in terms of their own experiences and context. Strategies are collectively grouped to attend to specific learning and teaching intent. These include:

- challenging students' existing ideas
- helping students to work out some of the 'scientific' explanations for themselves
- practising using and building the perceived usefulness of scientific models
- encouraging students to go further and identify phenomena not explained by the currently presented scientific model

- providing open problems to be explored via play or through problem-solving
- focusing students' attention on overlooked detail
- providing opportunities that promote reflection on and clarification of existing ideas (DEECD, 2006).

Considering the context

Developing deep knowledge and understandings of the social construction of scientific knowledge depends upon science learning experiences that, by purposeful design, expose students in an authentic way to a diverse range of experiences, including political, social and religious points of view. Of particular importance are experiences that highlight cultural understandings, which are different ways of seeing and knowing about the world. These experiences highlight not only ways of knowing that are different from Western science, but can be used to explicitly emphasise the facets of Western science that characterise it as a particular way of knowing the natural world (see Chapter 2 for a more in-depth exploration).

Formulating possible solutions to a range of problems that require students to consider the social, religious, political, cultural and ethical implications of their decisions can be a valuable way to demonstrate the interconnections between science and society. Through discussions and debates, interactions with community leaders and a variety of different cultural perspectives, students are more likely to develop understandings of social laws and the social and political expectations related to a variety of contentious socio-scientific issues. Case Study 5.1 speaks to these considerations by coming up with solutions for a large-scale social issue impacting largely tropical and sub-tropical areas of the developing world, such as Africa, Central and South America, Southern Asia and Oceania.

Learning experiences that also enable students to build relationships with scientists and experts relevant to their fields of study—whether through face-to-face interaction or the use of technology link-ups—can be very beneficial. Such experiences have the potential to not only enhance conceptual understandings, but also build a sense of purpose, importance and motivation for learning, and further a sense of ownership of student learning. They also expose students directly to the work of scientists and enable them to interact with the scientific community.

Primary and secondary school teachers can draw on community partnerships and contexts to welcome different ways of knowing science into science lessons and subsequently impact on student learning in different ways. This becomes particularly important when we consider the decline in the number of students taking up science studies in the senior years of secondary school and beyond, which is commonly connected with a general disenchantment and disengagement with how school science is presented. Most concerning—and perhaps this is where we can have the biggest impact—is the marked drop in student interest in science during the lower years of secondary school. These issues are underscored by students' perceptions that the science they learn in school is dull, decontextualised, difficult, teacher-driven and all about fact-focused content. In particular, science education in the primary years has an important role to play in fostering an enjoyment of and interest in science learning that can potentially set students up with a connection to science for life. This approach highlights the importance of supporting learning in ways that are aligned with our students' experiences and environment. Learning science in this way provides students with engaging and relevant concrete experiences that enable them to more clearly see the links between the science phenomena they are studying and their lives.

Fostering inclusive learning environments

Establishing learning environments that entrust students with their personal learning is essential for understanding science as a human endeavour. Students need opportunities to work autonomously, collaboratively and with wider community groups to develop a sense of authentic purpose for their science learning, while also being encouraged to notice themselves as learners and thinkers and to consider the types of interactions that are and are not productive. To achieve this, teachers can focus more on creating the most effective conditions for such learning as opposed to focusing on the content to be delivered. Such conditions must recognise that learning is not so much about digesting chunks of information, but involves developing valuable critical thinking, communication and problem-solving skills for lifelong learning. These are certainly the skills that underpinned the success experienced in Case Study 5.1. This is not to underplay the scientific knowledge required, but that alone would not have come up with the insights that the team did.

A possible framework that primary and secondary teachers can use to support quality learning in science is an approach known as Universal Design for Learning (UDL). UDL is an educational framework for curriculum development that guides the creation of flexible and responsive learning experiences that cater for individual learning differences and provide all students with opportunities to learn (Meyer, Rose & Gordon, 2014). This framework was designed to break down a significantly inhibiting and unintentional factor that can be a barrier for many learners: inflexible, one-size-fits-all curricula. Originally, UDL was considered for students considered to be on the margins of the mainstream population, such as those who were gifted and talented or had disabilities. These students are particularly vulnerable when variability in a classroom is not addressed. Now, however, it is seen as an approach that addresses variability by focusing on using flexible learning outcomes, strategies, resources and assessments to support the inclusion of all students. The following three principles guide UDL, and should be considered interdependently rather than independently to best meet students' learning needs:

- multiple means of representation, which provide students with different ways of acquiring information and knowledge
- multiple means of action and expression, which provide students with alternatives for demonstrating what they know
- multiple means of engagement, which tap into learners' interests, offer appropriate challenges and increase motivation.

What makes the UDL framework so important and effective in terms of inclusion is that, instead of focusing on a few students, it looks at ways of differentiating the learning experiences we plan and implement to suit the needs of *all* students.

Summing up

A response to the question 'What are the implications of NoS for science learning and teaching?' can be summarised as follows:

- The traditional purpose and approach to science curriculum and the teaching of science in schools have valued academic outcomes. In response, teachers have largely selected tasks that aim to achieve knowledge acquisition.
- For students, such experiences perpetuate a view of science as a rigid body of absolute unchanging truths consisting of isolated 'facts', devoid of human imagination and logical reasoning, disconnected from society and culture and having no real application to everyday life.
- Valuing and attending to the nature of science in school science necessitates rethinking the purpose and intention to science learning in terms of the importance of developing images of science that are consistent with current scientific practices.
- Science teaching is about acknowledging and attending to the importance of providing learning experiences that explicitly emphasise creative and critical thinking, observation,

interrogation of data and evidence, and social interaction in the construction of knowledge, highlighting the influence of human endeavour.
- Developing deep knowledge and understandings of the social construction of scientific knowledge depends upon science learning experiences that, by deliberate design, expose students in an authentic way to a diverse range of experiences (e.g. culturally, politically) while also exploring ideas and concepts situated in contexts beyond the classroom.
- The learning environment has a significant impact on student learning, and this requires teachers to move beyond content and consider how to foster the conditions required to support the development of valued student learning behaviours.

This section explored the implication of NoS for science learning and teaching by considering how classroom practices can shift from emphasising school science to reflecting contemporary approaches to science. This is important if we, as science educators, are to make changes to our own ways of thinking about science, so that it can trickle down and be realised in our teaching approaches and the conditions we foster to enhance learning. Using NoS as a lens opens up an opportunity for science in schools to be authentic, relevant and reflective of how science is actually practised.

CONCLUSION: RISING TO THE CHALLENGE

The challenge for teachers at all levels is to determine how NoS can be explored and enacted effectively through engaging with the contemporary world of science. In particular, this chapter encouraged you to consider what this might look like in terms of classroom practice and how it might be demonstrated in student learning. This requires a transformation of school science pedagogy, which is not an easy shift to make. Teachers themselves are as much a product of the system structures and agencies to which they are accountable as students themselves. As professionals, primary and secondary school teachers work within a society that says it values the need to understand NoS to develop more science-savvy citizens, yet existing systems and assessment requirements attempt to identify learning based on norms that are formed analytically, but not using empirical evidence. Current assessment approaches overwhelmingly value the reproduction of factual information as the best measure of a knowledge and understanding of science. As a result of this emphasis, teachers lack a good understanding of NoS, and this limits their ability to effectively engage students in rich discussions interrogating the social construction of scientific knowledge. There still exists a dominating tendency to dispense knowledge. This needs to change through science teachers becoming aware of how to facilitate the conditions under which students are more likely to construct meaningful understandings, critically evaluate scientific evidence and apply scientific knowledge in meaningful contexts.

DISCUSSION QUESTIONS

5.1 How would you define science? What differences do you think there might be in how you defined science prior to reading this chapter?

5.2 What are some of the key actions and principles that define and characterise the development of NoS?

5.3 In a world where *fake news* and *alternative facts* are prominent features in the information we consume, how might you as a teacher of science (drawing on your knowledge of NoS) support your students in navigating these challenges to their learning of science?

5.4 Identify and describe three ways in which you could integrate NoS into your approach to science learning and teaching?

5.5 What are some strategies that you might adopt to ensure that your understandings and knowledge of NoS, both as a concept and in practice, remain cutting edge and up-to-date?

FURTHER READING

Corrigan, D., Dillon, J. & Gunstone, R. (2007). *The Re-emergence of Values in Science Education*. Rotterdam: Sense.

Corrigan, D. & Smith, K. (2015). 'The role of values in science education'. In J. Deppler, T. Loreman, R. Smith & L. Florian (eds), *Inclusive Pedagogy in the Curriculum*. Bingley: Emerald Books, pp. 99–117.

Rennie, L. (2011). 'Blurring the boundary between the classroom and the community: challenges for teachers' professional knowledge'. In D. Corrigan, J. Dillon & R. Gunstone (eds), *The Professional Knowledge Base of Science Teaching*. Dordrecht: Springer, pp. 13–30.

—— (2015). 'Making science beyond the classroom accessible to students'. In D. Corrigan, C. Buntting, J. Dillon, A. Jones & R. Gunstone (eds), *The Future in Learning Science: What's in it for the Learner.* Dordrecht: Springer, pp. 151–74.

REFERENCES

Akindehin, F. (1988). 'Effect of an instructional package on preservice science teachers' understanding of the nature of science and acquisition of science-related attitudes'. *Science Education*, 72, 73–82.

Bell, R. (2001). 'Implicit instruction in technology integration and the nature of science: There's no such thing as a free lunch'. *Contemporary Issues in Technology and Teacher Education,* 1(4). Retrieved from www.citejournal.org/vol1/iss4/currentissues/science/article2.htm.

Corrigan, D.J. (2015). 'Values and learning science'. In R. Gunstone (ed.), *Encylopaedia of Science Education*. Dordrecht: Springer, pp. 1091–3.

Department of Education and Early Childhood Development (DEECD) (2006). *Science Continuum P–10*. Canberra: DEECD.

Doherty, P. (2005). *The Beginner's Guide to Winning the Nobel Prize: A Life in Science*. Melbourne: Miegunyah Press.

Duschl, R. (2000). 'Making the nature of science explicit'. In R. Millar, J. Leach & J. Osborne (eds), *Improving Science Education: The Contribution of Research.* Buckingham: Open University Press, pp. 187–206.

Fensham, P. (1988). 'Familiar but different: Some dilemmas and new directions in science education'. In P. Fensham (ed.), *Development and Dilemmas in Science Education*. London: Falmer Press, pp. 1–26.

Hildebrand, G. (2007). 'Diversity, values and the science curriculum'. In D. Corrigan, J. Dillon & R. Gunstone (eds), *The Re-emergence of Values in Science Education.* Rotterdam: Sense, pp. 45–60.

Hodson, D. (2003). 'Time for action: Science education for an alternative future'. *International Journal of Science Education*, 25(6), 645–70.

Lederman, N.G. (1996). 'Syntax of nature of science within inquiry and science instruction'. In L.B. Flick & N.G. Lederman (eds), *Scientific Inquiry and Nature of Science, Implications for Teaching, Learning, and Teacher Education*. Dordrecht: Springer, pp. 301–18.

Meyer, A., Rose, D. and Gordon, D. (2014). *Universal Design for Learning: Theory and Practice*. Wakefield, MA: Center for Applied Special Technology.

National Research Council (1996). *National Science Education Standards*. Washington, DC: National Academies Press.

Nowotny, H. (2005). 'The changing nature of public science'. In H. Nowotny, D. Pestre, E. Scmidt-Abmann, H. Shulze-Fielitz & H. Trute (eds), *The Public Nature of Science Under Assault: Politics, Markets, Science and the Law*. Berlin: Springer, pp. 1–27.

Osborne, J.F., Ratcliffe, M., Collins, S., Millar, R. & Duschl, R. (2003). 'What "ideas about science" should be taught in school science? A Delphi Study of the "expert" community'. *Journal of Research in Science Teaching*, 40(7), 692–720.

Schwartz, R., Lederman, N.G. & Crawford, B. (2003). 'Developing views of nature of science in an authentic context: An explicit approach to bridging the gap between nature of science and scientific inquiry'. *Science Teacher Education*, 88(4), 610–45.

CHAPTER 6
SCIENCE INVESTIGATIONS

DEBORAH CORRIGAN, ANGELA FITZGERALD *and* **GILLIAN KIDMAN**

LEARNING OUTCOMES
After reading this chapter, you should be able to:

- think differently about what approaches and processes might contribute to understandings of a science investigation
- demonstrate an awareness of the range of activities that inform the structure of an investigation
- consider the pedagogical approaches that can be used to promote and enhance student learning in science of and from the investigative process.

This chapter explores the different approaches that science investigations can take so that, as teachers of science, we can think about how we can authentically represent these types of scientific work to our students. Once again, we will start with a real case study of a scientist at work, and use this to highlight the important elements of science investigations. It is important to remember why scientific investigations are important. All scientists are constantly trying to improve their understanding and explanation of the natural world. In pursuing this goal, scientists agree that there are a number of rules connected with the scientific process:

- use empirical data
- use reasoning in constructing explanations
- avoid bias
- present explanations for sceptical review (by peers).

One of the authors of this chapter (Kidman) was a scientist working on the project briefly outlined in case study 6.1. This experience offers unique insights, which will be shared here, into the process that drove this particular investigation.

The beginning point for this project was a set of questions—for example, 'What soil type might foster aggressive invasive growth of *Arachis glabrata*?' Next, decisions were made about how best to investigate these questions. In response to the example question, we would establish test plots of clay, loam, sand and combination soils upon which to grow *A. glabrata*, as well as look at what others have done before. This would be done through a **literature review**. If the literature review produced new information—that is, if another scientist had conducted similar research—the researcher could, if necessary, refine the question—for example, a definition of aggressive invasion.

literature review a process whereby researchers review work that has already been published in this field in order to inform and progress the current research project. The literature is sourced from academic journals and some academic books

The methodology would constantly be reworked as the research questions evolved and changed. Finally, the process of taking samples, deciding what to measure and

CASE STUDY 6.1: CSIRO project—investigating the growth patterns of *Arachis glabrata*

Context

A. glabrata (Figure 6.1) is a creeping peanut native to South America. It has a high nutrient value, tolerates high-density beef cattle grazing and is highly resistant to pests and disease. As an introduced species to Australia, the close monitoring of its growth habits and spread was important to gauge its impact on the Australian environment.

Figure 6.1 Flowers and foliage of *A. glabrata*
Source: http://www.tropicalforages.info/key/Forages/Media/Html/Arachis_glabrata.htm.

The investigation needed to design experiments that involved various soil types, differing grazing rates, monitor environmental conditions and measure the growth and spread of the plant (see Vos & Jones, 1986 for more details).

making measurements would begin—for example, controlling the number of plants initially sown on each plot and making consistent growth measurements. After this, a return to and consideration of the questions would again be needed to gauge whether the appropriate information was being collected and the method of investigation was working as required. In doing this, modifications would be made as necessary to such aspects as what to collect, what to measure and techniques to be used and data collected. It would be useful to return to the literature review to see how others had represented their data, then the data would be represented in a conventional manner for communication to the public. This communication might take the form of tables, line graphs and scientific diagrams. The representation of data and any analysis would be considered in relation to the research question. A subsequent question would tend to be asked at this point: Is there a possible solution? If so, there would be a return to the starting point to engage in this investigative process again to check that the data and possible solution were not simply an accident.

This is definitely not a **linear practice**; rather, it is an **iterative process**, which involves continuously cycling back to the research question. The only certainty in this approach is that it starts with a question or hypothesis, or an interesting phenomenon. What happens after this point is up to the scientist and determined by the nature of the question and the chosen methodological process.

linear practice a process where something progresses directly from one stage to the next, with a start and end point

iterative process a process where a decision or result is reached by repeating rounds or cycles of analysis

Case Study 6.1 explored the approach that one team of scientists took to a problem that needed addressing. Case Study 6.2 describes the undertaking of a science investigation in a school context.

In this case study, a somewhat different investigative approach was taken, although

CASE STUDY 6.2: Is it real science?

I had decided that even in Term 4 it was not too late for students to take control and to have input into the direction of their own learning. On the bench, I had collected a few items that are not normally seen in the lab, such as nappies, detergents, chocolates and shampoo.

'What is consumer science?' I asked the class.

'Is it a real science?' Nicole asked a little hesitantly.

'Is it part of chemistry, physics or biology?' asked Emma.

Michele, my over-achiever, said, 'It's all about testing all those consumer products to see whether they are what they claim to be.'

This began the discussion about how scientific testing can be used to investigate the validity of claims made by companies about their consumer products. Students began to share stories about hair products, cleaners, nail polish removers and makeup that costs a fortune but in the end did not deliver the quality.

I felt that for this investigation I would need to commit and give a reasonable amount of time for students to discuss and plan their method. I wanted them to own their own ideas and to take responsibility for the direction of their own investigation. They were formulating their ideas and in some cases arguing about the best method for their investigation.

'That's not going to work,' said Anne. 'You're not testing just one thing.'

'Should we test each product at least three times to get more accurate results?' asked Ashley. 'What about a control set up?'

'How are we going to collect and show our results?'

There was a sense of excitement and the buzz in the room elevated to a small roar. The students were all testing different products using different techniques, and there was a real sense of purpose to their work. We had nappies being tested for absorbency, melting chocolate brands being timed for melting times and diet soft drinks being put to the test to see whether they were indeed sugar free. These were just a few examples. The students were all keen to discover their own answers, repeating their tests to validate the accuracy of their data and engaging in vigorous discussion about the results they were finding.

They were making progressive recordings, photographing their results and working in a collaborative manner, with each person assigned a role. There was no one who was without a task to perform.

'The cheaper brands do the job just as well, if not better, than the most expensive brands,' said Mary.

'What a rip off!' said Ashley. 'I'm telling Mum that her brand of detergent froths less than the home brands and costs three times as much.'

'Is it over? Can we test another product?' implored Ashley. 'I can't believe how much fun this was—was it really science?'

Source: Tripaydonis (2008), pp. 85–7.

a number of similar steps were evident. For example, while there were some questions generated, the approaches that the participants made explicit included:

- using scientific testing to investigate the validity of their claims
- articulating what was the best method for their investigation
- testing different products using different techniques
- discovering their own answers
- repeating their tests to validate the accuracy of their data
- vigorous discussion about the results they were finding
- making progressive recordings by photographing their results
- working in a collaborative manner, with each person assigned a specific role.

These two cases highlight that scientific investigations can be carried out in a variety of ways. As Wong and Hodson (2010) remind us, it is the *context* that determines the method of an investigation. While Case Study 6.1 relied on the generation of questions, Case Study 6.2 was focused on investigating the **validity** of **claims** made about consumer products. The essential start to any scientific investigation is to consider whether the question, claim or hypothesis driving the work can be investigated in a scientific manner. For example, it is difficult to scientifically investigate the question 'Does God exist?', as such an investigation would not enable the use of an empirical approach, which is fundamental to the science investigation process. As Gott and colleagues (n.d.) point out:

validity the extent to which measurements or assessments measure or assess what they are supposed to

claim a statement that states or asserts that something is the case

> Investigations must be approached with a critical eye. What sort of link is to be established, with what level of measurement and how will opinion and data be weighed as evidence? This pervades the entire scheme and sets the context in which all that follows needs to be judged.

These two case studies highlight that many possible investigative approaches exist. In this chapter, we explore a range of investigative approaches used in science as well as highlight some of the important ideas that underpin the investigative work done by scientists. In addition, we share a range of pedagogical approaches that can be used to support students in the primary and secondary years to develop their science investigative skills.

WHAT CONSTITUTES A SCIENCE INVESTIGATION?

Any scientist will tell you that there is no one way to approach scientific investigations. A scientist will use a variety of approaches or pathways, depending first on what question they are asking, what funding they may have for equipment, what is ethical and so on. However, many textbooks used for school science begin with a section called 'the scientific method'. This highly stylised template for conducting scientific investigations is largely a myth that has been created to make science investigations simple for students of all ages to follow. In reality, the scientific method has done more damage to students developing appropriate approaches to scientific investigation than anything else, as it does not account for the context in which such investigations occur. For example, Wong and Hodson (2010) remind us that context determines the method of an investigation, not the sequence of steps you remembered above. They point out that:

> It is common for a page or two to be set aside for a description of 'what scientists do', usually presented in terms of a sequence such as observation, generalization, hypothesis, experiment, and theory. The notion of a fixed, all-purpose approach is then reinforced by dozens of laboratory reports in which students are required to emphasize each step of 'the scientific method'.

Harwood (2004, p. 29), like many others, criticised the scientific method model as representing the processes of science as only being 'conducted by formulating a rigorous hypothesis and then testing it'. More importantly, he claimed that this scientific method model implied that the 'process of science is dependent on following a proscribed set of steps' (2004, p. 29). Following on from this, based on interviews and focus group discussions with academic research scientists, Harwood (2004, p. 30) identified nine activities that extend far beyond the scientific method model and that 'scientists engage in as often as necessary and move among activities in a pattern dictated by their specific needs'.

These nine activities provide a structure for science investigation that does not represent a stepwise process and is not presented in the scientific method model. They also provide insights into the process that both primary and secondary teachers can draw upon to inform their practice and support their students in learning about the investigative process from a scientific perspective. These activities range from identifying problems through to communication, and will now be described in further detail. It should be noted that some activities are discussed in more detail than others, which reflects their connection and relevance to science learning and teaching, along with the purpose of this chapter, which is to help prepare you for the role and place of science investigations in learning science.

variable an element, feature or factor that is liable to change or vary

Problem definition

> An investigation is an attempt to determine the relationship, or lack of one, between the **independent** and **dependent variables** or between two or more sets of data. Investigations take many forms. (Gott et al., n.d.)

independent variable the element, feature or factor that the investigator can change—for example, time is often the independent variable as the investigator can decide what time intervals they will use

dependent variable the element, feature or factor that changes in value in line with the change in the independent variable—for example, in an investigation of the weather, air temperature (the dependent variable) is measured at particular times (the independent variable)

Defining a problem may require a single investigation or a series of investigations to work out how to reach a solution. Where a series of investigations is required, these may be nested within a wider or broader approach to an issue. As Gott and colleagues (n.d.) point out, investigations attempt to determine whether a relationship exists. One way to do this is to consider variables—those elements, features or factors whose value is likely to change or vary. Other problems can be more descriptive in nature, such as describing environments straight after a bushfire and returning some time later (weekly or monthly) to make the same observations again.

Posing questions

While the type of questions that are used to frame a scientific investigation can be quite broad, they do need to be questions that can be explored in a scientific way. This means that they will need to have data generated from empirical processes and attempt to avoid bias. As pointed out in Chapter 5, because scientists conduct these investigations, they bring with them their own knowledge, beliefs, opinions and experiences. The need to try to reduce the effect of these personal characteristics is an important part of the scientific process. One way to reduce such bias is to use processes that other scientists have used in lots of different settings—as these will generally mean that such personal bias has been minimised. The same can apply to instruments: those instruments that have been calibrated prior to use and have produced consistent readings regularly will reduce instrument bias. There are many other ways to reduce bias. Data generated should be *reliable*, in that the data are consistent, and *valid*, in that the data are measured in ways that measure what was intended, to provide the evidence necessary for the question to be answered.

The following questions could be used to decide what type of investigation is appropriate for your particular purpose as highlighted in Table 6.1, which provides an explanation of what characterises each type of investigation. Each investigation type can be used in both primary and secondary school settings. Remember that, as teachers, you should engage in science investigations with a variety of purposes so your students develop the full range of investigative skills.

causal relationship when the change in one *variable* is the result of change in the other variable

control group the group in an experiment that does not receive any treatment or experience any change; used as a benchmark (or baseline) for measuring purposes

1 Do you want to simply draw some conclusions about your selected topic? If YES, then you want to undertake a *descriptive investigation.* If NO, go to Question 2.
2 Do you want to determine relationships between variables? If YES, then you want to undertake a *comparative investigation.* If NO, go to Question 3.
3 Do you want to determine a **causal relationship** between variables? If YES, then you want to undertake an *experimental investigation.* If NO, go back to Question 1 and reconsider the intended purpose as you have missed the most appropriate investigation type.

Table 6.1 Characteristics of investigation types

	Involves a hypothesis	Involves variables	Involves a control	Keywords
Descriptive e.g. making measurements and describing observations	No	No	No	Observe, describe, list, identify
Comparative e.g. comparing measurements, comparing observations	Yes	Yes	No	Compare/ contrast, similarity/ difference, categorise
Experimental e.g. measuring germination with varying moisture	Yes	Yes	Yes	**Control group,** experimental group, constants

The differences between the various types of investigation can be further explained if we consider the types of questions that might drive each process. For example, if you were teaching a topic on healthy eating, digestive systems or energy and metabolism, the questions in Table 6.2 would all be valid.

Investigating what is already known

Generally, the results of any investigation are compared with other data. Using secondary data—data collected by others—can provide a valuable source of additional evidence. Additionally, **triangulation** (comparison with other methods and/or approaches) can also strengthen the validity of the evidence that has been gathered.

triangulation a technique that facilitates the validity of data through verifying the data across two or more sources. It often requires the use and combination of several research approaches in the study of the same phenomenon

Observing

Observation of objects and events can lead to informed description and the generation of questions

Table 6.2 Examples of differing investigation questions

Descriptive investigation questions	Comparative investigation questions	Experimental investigation questions
What is the recommended weekly calorie intake for Australian children/teenagers?	What is the difference between Australian male and female children/teenagers who exceed the weekly calorie recommendations?	What is the relationship between Australian male and female children/teenagers' weekly calorie intake and time spent doing physical activities?
What happens when M&Ms are put into water?	Which M&M colour dissolves the fastest?	How does water temperature affect the dissolving of M&Ms?

> to investigate further. Observation is one of the key links between the 'real world' and the abstract ideas of science. Observation, in our definition, does not include 'measurement', but rather deals with the way we see objects and events through the prism of our understanding of the underlying substantive conceptual structures of science. (Gott et al., n.d.)

Observation can take a number of forms, such as:

- observing objects or events using a key or a **taxonomy**
- capturing what was witnessed by drawing a map or diagram (e.g. fieldwork).

The process of observing and what is learnt through this act can in itself prompt an investigation.

taxonomy a scheme of classification—for example, a taxonomy of mammals

Articulating the expectation

Investigations need to distinguish between opinion based on scientific evidence and ideas, and opinion based on non-scientific ideas, such as prejudice and hearsay. Scientific investigations, as stated previously, seek to establish links (and the forms of those links) between two or more variables. Such links can be *causal*, where a change in one variable causes a change in another, or *associative*, where changes in one variable and changes in another are linked to a third or more variables. These investigations also need to account for the types of measurements to be made. For example, *interval data* that measure a continuous variable (such as length, which can have any numerical value) are more powerful than *ordinal data* (data that often have descriptions such as very small, small, medium and so on, and that must be given a rank ordering), which in turn are more powerful than *categoric data* (a label such as 'type of metal' is the variable and the values are iron, copper, tin and so on). Additionally, the investigation may be *nested*, where the measurements taken in one investigation only become meaningful when set within a wider investigation. An example of investigations with nested or multiple variables would be the effect of width and length on the strength of a bridge.

Doing an investigation: Processes and types

The process of doing or conducting an investigation often involves understanding the underlying relationships between variables, which can require the use of an instrument to make a measurement. For example, a measuring cylinder (the instrument) converts volume (the variable) to a length of liquid in a column (the measurement). In this instance, volume is measured by length and multiplied by a known area (the cross-section of the cylinder) and the graduations marked on the cylinder—for example, 100 millilitres represents this calculation. Similarly, a thermometer converts a change in temperature first into volume and then into a change in length of a thread of mercury. Some instruments convert the variable (e.g. speed) to an angle on a curved scale (e.g. a speedometer). In other instances, the notion of 'instrument' might not be easily recognisable, as when lichen is used as an indication of pollution due to its ability to derive water and essential nutrients mainly from the atmosphere, or pH paper is used as a measure of the amount of chemical change in an acid–base reaction.

Instruments also rely on more complex and less direct relationships, which can be of a range of types, such as linear (e.g. thermometer as in measuring a length of mercury), non-linear (pH as in measuring concentration in a logarithm—base 10–scale), complex (lichen can also indicate other environmental factors such as air movement) and multiple (such as in medical diagnosis where multiple instruments such as x-rays and magnetic resonance instruments [MRIs] are used), and often indirect relationships, such as x-rays approximating bone density, are employed.

In doing or conducting an investigation, there is also a range of approaches that can be drawn upon with some having already been discussed. But there are many more possibilities. Another major type of scientific investigation that hasn't been explored in this chapter yet is fieldwork, which is commonly used in relation to biological and earth sciences.

Fieldwork is an essential component of science education in primary and secondary classrooms. Below is a selection of fieldwork techniques that can be used in both settings. Extensive detail has been omitted, as a plethora of information is available. If you are wanting to conduct an investigation involving fieldwork in naturalistic settings, then the excellent website of the Royal Geographical Society (n.d.) is a great place to look for ideas and technique. There is an extensive Schools and Education section with excellent coverage of fieldwork techniques.

1 *Field sketching and photography* are fundamental parts of any fieldwork investigation. They constitute a simple qualitative technique that, done correctly, will support the data collected and enhance the whole investigation. For example, if you are mapping the flora in a particular area, photograph the ground so that the colour, leaf shape, height of the plant, type of flowers and so on can be captured. Sketching is best done using a soft 2B or 3B pencil. There should be a purpose for what is being sketched or photographed. Ensure your students include details such as date, location (GPS coordinates), title, direction of view, measurement/scale and, most importantly, the focus of the sketch/photograph as this will align with the aims of the investigation.

random sampling a sampling technique that results in each sample having an equal chance of being chosen

2 *Sampling techniques*
- **Random sampling** can be used where there is a large population from which to select. It avoids bias in selection.
- *Systematic sampling* uses uniform intervals at which to sample. For example, every fifth person is approached or every 50 centimetres of water is sampled. The chance of bias is greater as not all people (no people who are the first, second, third or fourth people are chosen) or sites (such as at 25 or 60 centimetres depth) are considered.
- Stratified sampling can happen in two ways. A *stratified systematic sample* is where an entire population is subdivided into smaller groups, then each of these groups is systematically sampled. A *stratified random sample* is where an entire population is subdivided into smaller groups, then each of these groups is randomly sampled.

3 *Transects* or a straight line or narrow section through an object or natural feature like a river are intended to investigate the structure or shape of natural features such as a landscape. This is usually done in conjunction with the above sampling techniques. The convention when conducting transects is to use a straight line for the sampling.

4 *Water safety and quality* require close attention when students are involved, particularly when judging whether water is suitable for drinking. You should not be deterred by safety requirements, but do be aware that extra vigilance is necessary. The customary water quality measurements are temperature (and how it varies with depth), dissolved substances such as oxygen, nitrates and phosphates, water turbidity (or the cloudiness of the water), water pH and microscopic species (such as bacteria) abundance.

5 *Cross-sections* are normally associated with streams and are a sketch that shows the stream width and depth, drawn to scale on graph paper.

6 *Investigating opinions*

- *Surveys* tend to be open-ended (rather than closed) questions, and provide an opportunity for respondents to give their opinions. For example, a closed survey question could be 'What is your age?' An open-ended survey question could be 'What do you believe caused the water cloudiness?'
- *Questionnaires* are made up of questions and often accompanied by boxes of response options. Options might be ticks in a box, a rating between 1 and 5, or circling a response.

Examining the results

It is important for you and your students to appreciate that the gaining of skills in statistics and data analysis is as much a part of doing an investigation as the work in generating the data. As a teacher, you need aim to help your students to link the use of statistical techniques to the acquisition of scientific knowledge and the method of inquiry.

Descriptive statistics (mean, median, mode, range)

'Descriptive statistics' is the term used to summarise the raw data collected in an investigation. Its full use is not recommended for students in the lower primary school, as they will not have developed the necessary mathematical understandings for mean and median. However, young children certainly can analyse modal statistics, or what is the most common data value.

Central tendency of data is a common type of descriptive statistic, and is an estimate of the centre of the numerical values. The three most common measures of central tendency are mean, median and mode.

The *mean*, or average, is the most commonly used form of central tendency. To calculate the mean, simply add up all the values and divide by the number of values. For example, the mean or average number of pendulum swings dropped at a fixed height score is determined by summing all the swings and dividing by the number of drops. For example, consider the following pendulum data: 21, 22, 28, 26, 28, 15, 25, 21. The sum of these eight values is 186, so the mean is 186 ÷ 8 = 23.25.

The *median* is the score found at the middle of the set of values. To compute the median score, you need to list the values in order of size—in numerical order, (15, 21, 21, 22, 25, 26, 28, 28)—and then locate the score at the mid-point of the sample. We have eight scores so scores 4 and 5 represent the halfway point. We need to find the mid-point between 22 and 25. So the median number of swings would be 23.5.

The *mode* is the number of swings that occurs the most frequently. To determine the mode, you order the scores as we did for the median, then count each one. The most frequently occurring value is the mode. In this pendulum swing example, the scores 21 and 28 both occur twice, so we can say our data has a bimodal distribution ('bi' meaning two).

The *range* is a measure of dispersion or spread, and is the highest value minus the lowest value. In our pendulum data, the highest number of swings is 28 and the lowest is 15, so the range is 28 – 15 = 13.

It can be confusing to know when to use which type of measure of central tendency, but basically it needs to fit the purpose. For example, the mean takes account of all the pendulum swings, whereas the median takes only those swings in the middle of the range of swings and the mode only the most common number of swings.

Controls and variables

Consider an investigation using the school's vegetable patch to grow tomatoes. You and your students will need to fertilise the tomato plants as they grow, so you collectively need to decide which fertiliser to use. You prepare two sections of the vegetable patch as trial plots for your tomato plants. Each of the two sections is divided into four, with each part treated with a different fertiliser to see which produces the greatest tomato yield. You need to control as many variables as possible—such as irrigation, soil type and tomato variety—and only vary the fertiliser. This means any changes in fruit yield can be attributed to the fertiliser.

In this investigation, there are three common types of variables: the independent, the dependent and the controlled. The independent variable can be set and varied by you and the students; this means the independent variable will be the type of fertiliser used. The dependent variable changes when the independent variable changes, so the dependent variable in this example is the tomato yield. There are also controlled variables, which are not tested, but are kept the same (*controlled*). In this example, the quantity of fertiliser used in each section was kept the same—it was controlled.

There is a relationship between the variables and how they are reported in a table (see Figure 6.2). The independent and dependent variables always go in the same positions in a table. The independent

Fertiliser type	Tomato yield (number per section)		
	Tomato yield section A	Tomato yield section B	Total tomato yield (Section A + Section B)
Type A			
Type B			
Type C			
Type D			

Independent variable

Dependent variable

Figure 6.2 Positions of variables in a table

variable is conventionally placed in the left-hand column. The dependent variable is written at the top of each column to the right of the independent variable.

Tables and graphs

Tables and graphs are visual representations of your data, and are used to show patterns and relationships. However, you and your students will need to be able to select the most appropriate table or graph for your data. The type of data you and your students collect determines the sort of table and graph you can use.

There are two main types of data: categorical and numeric. If you are collecting *categorical data*, the values and observations can be sorted into groups or categories, and bar graphs and pie graphs can be used. With categorical data, you can rank your data points, or put them into order, so the term used is *ordinal.* Ordinal sounds like 'order'—it is the order that is important. Examples of ordinal data are typically non-numeric concepts like satisfaction, happiness and discomfort, which can be expressed on a rating scale.

Numeric data are measured values. They are placed in ascending or descending order, and are either discrete or continuous. Discrete values are counted and so are a particular number. Examples are the number of red-haired students in your science class, or the number of heartbeats in 30 seconds. With continuous data, you make measurements, so data is any value in a range—a person's height or the time taken to walk to school.

The following is a summary of the types and uses of tables, data and graphs:

- *frequency table*—categorical or discrete numerical data
- *grouped frequency table*—numerical data
- *picture graphs* (pictographs)—categorical data
- *bar graphs* (vertical/column or horizontal bar graphs)—categorical and discrete, ungrouped numerical data; used

to compare between different groups or to track significant changes over time

- *line graphs*—continuous data; used to track changes over time; also used to compare changes over time for more than one group
- *dot plots* (scatter plots)—categorical and discrete numerical data
- *pie graphs*—categorical and discrete numerical data; best to use when you are trying to compare parts of a whole
- *histograms*—ungrouped discrete and continuous numerical data.

It is essential for all teachers to be able to create and read graphs. Not only will you develop these skills in your students, but you will also be required to understand student data of achievement, such as interpreting NAPLAN data.

Reflecting on the findings

Any evidence from science investigations must be considered in the light of personal and social experiences of the investigators. When presented with evidence about which you need to make a judgement, a range of factors needs to be considered. Gott and colleagues (n.d.) have identified several points for consideration:

- *Credibility of the evidence.* Is the evidence consistent with conventional ideas, common sense and personal experience? Credibility can also be influenced by the type of evidence being presented—for example, statistical versus anecdotal (from a personal story or hearsay) evidence. For instance, when buying a toaster, many people might rely on testing done by a consumer organisation such as Choice, which presents the findings from tests (e.g. how many times the toast was burnt—statistical in nature) as opposed to people's reviews of the toaster (anecdotal in nature).

- *Practicality of consequences.* Are the implications of the evidence practical and cost-effective?
- *Investigator bias.* Is there evidence of inherent bias resulting from funding sources, intellectual rigidity or allegiance to a particular ideology, or related to who benefits and who is disadvantaged by the evidence? For example, some research presented on the benefits of taking vitamin supplements may be funded by a vitamin company. The debates around climate change and evolution are examples of the influence of different ideologies.
- *Power structures.* Evidence may be considered more favourably if there is more influence in play from certain groups, such as politics, influential bodies or individuals, or special-interest groups.
- *Paradigms of practice.* Different investigators may work within different **paradigms** (which are a distinct set of concepts or thought patterns), such as the difference between engineers and scientists, who both learn science but apply it in very different ways. The evidence may take on a different status or meanings in these different paradigms, depending on which paradigm is seen as the most expert in a given context. For example, in determining where forces exist, scientists may be seen as more expert, whereas in using forces for creating structures, engineers would be seen as more expert.

paradigm a framework containing the basic assumptions, ways of thinking and methodology that are commonly accepted by members of a scientific community

- *Acceptability of consequences.* Evidence can be illogically denied or dismissed, based on public or political fear of its consequences. Prejudice and preconceptions may play a part in this.
- *Status of investigators.* The academic or professional status, experience and authority of the investigators may influence the value placed on the evidence.
- *Validity of conclusions.* Conclusions should be limited to (and

not extend beyond) the available data. For example, in measuring the temperatures in a day, you could conclude whether the day was average or not by comparing it with past temperature data. This would be a valid conclusion. It would not be valid to conclude anything about the weather tomorrow.

While these areas of consideration may require higher levels of critical thinking, it is important to be aware of them so you can enrich your students' ways of knowing and doing science in general and science investigations in particular.

Communicating with others

The results of investigations need to published and subjected to peer and public review. However, findings should also be communicated to a range of audiences, and hence the communication tools and genres used need to change to suit the context. Corrigan (2015) has developed a schema for different types of science communication engaged in by contemporary scientists. This approach is seen as innovative, as it tries to distinguish between the broad areas of complex cognitive engagement needed for effective communication with different audiences for different purposes. The schema outlined below attempts to identify five areas of science-cognitive engagement that scientists, technologists and researchers are likely to encompass in their work.

Schema for science-cognitive engagement

1 *General public engagement.* This is probably the most basic level of communication; however, even though the sophistication of the science knowledge exchanged is likely to be quite elementary, challenges remain. Looking to effectively communicate insights into big ideas or complex processes using powerful metaphors or analogies is a creative and often demanding task that confronts many educators on a daily basis. Predictably,

not all scientists are skilled at communicating with the general public, which makes those who are—like Tim Flannery, Richard Dawkins and Brian Cox—highly sought after by both the mainstream media and the general public.

2 *Informed engagement.* This describes engagement by those who are conversant with a scientific field or discipline. They are informed and seek opportunities to share and improve their knowledge and understanding among competent peers with similar interests or expertise. This form of engagement is practised by everyone, from amateur interest groups and student societies to professional institutes and associations—for example, Soil Science Australia, the Royal Australian Chemical Institute and the Australian Academy of Science.
3 *Applied engagement.* This describes a broad engagement by scientists, engineers, technical designers and science communicators, who apply current scientific knowledge to develop real-world applications of technology or provide insights into fundamental processes of science. Their interests may include fields such as engineering, medical imaging, robotics or nanotechnology. They use the knowledge of science and its processes—for example, experimental design, analysis of data and modelling—to test and improve technology and its applications.
4 *Focused engagement.* This includes engagement that deals with the routine practicalities of communication practices within and between scientific or industry research centres. Examples could include system approaches for regular reporting on project challenges and achievements to project personal, routine laboratory meetings, initiatives exploring workflow or communication practices and team reviews of technical protocols. It could also include project reporting to industry and government, mentoring practices and career-building

and management within a specific research field or scientific organisation.

5 *Expert engagement.* This engagement involves science discipline authorities or research leaders acknowledged by their peers as experts and visionaries—for example, Nobel Laureates, or winners of the Prime Minister's Prize for Science, Eureka Prizes, AIP awards and AAS awards. This could include expert analysis or commentaries on new technology or recent scientific research discoveries, and their likely societal or cross-disciplinary impact. Experts regularly provide keynote addresses at conferences, and their insightful presentations and critical analysis are regularly sought by industry, field specialists and the general media.

In summing up, Harwood (2004) describes investigations as an iterative process that attempts to capture the constant movement between these nine activities as determined by the context. There needs to be constant checking of the appropriateness of the investigations through question(s) and the methods used to investigate, analyse and communicate them.

PEDAGOGICAL APPROACHES TO SCIENCE INVESTIGATIONS

When students engage in science investigations (or practical work) in schools, they often do not know what to do or look for, and even less often know what and how to communicate their work. Additionally, the equipment they use may be unfamiliar, which can impede the real thinking that students need to engage with if they also have to work out what to use and how it works. So it is important to find ways to assist students to undertake meaningful science investigations where the focus is on real thinking and learning rather than following a set routine.

Restructuring a science investigation

For many students in lower secondary school, their introduction to science investigations is often through recipe-style practical investigation, where students complete steps in a particular order, without any real consideration of why. In Case Study 6.3, a common concept such as density, which lower secondary students explore, has been restructured into an open-ended investigation where the students need to work collaboratively to find out the density of some common substances. For example, students can be given aluminium foil as the material they will use to try to find out the density of aluminium. In undertaking this investigation, students will need to think about how they will conduct this investigation and what equipment they will need in advance. The example provides some possible equipment, but this is just an incomplete guide to help in their thinking.

Another of the substances that students will need to find the density of is air. This can be quite problematic, and many students think air has no mass, so could not really have a density. To push the

CASE STUDY 6.3: Properties of substances—density

Aim

To find the density of some common substances.

Method

You have to devise a suitable method to find the density of the following substances. Some suggested equipment is provided (Table 6.3).

Table 6.3 Substances provided for density experiment

Substance	Given
Water	Water
Lead	Lead shot
Iron	Iron nails
Cork	Some cork stoppers
Pine wood	A pine joist
Air	
Oil	Cooking oil
Aluminium	Alfoil
Carbon dioxide	A flask of carbon dioxide
Rock (basalt)	Screenings

Possible equipment
Hammer, saw, measuring cylinders, top loading balance, rulers, pins, vacuum pump, tubing and clamps, balloons, beakers, overflow cans.

You have to decide IN ADVANCE what to do.

Make a table with your proposed method for each substance.

Substance	Proposed method
X	
Y	
Etc.	

Write up

Use the following format:

Substance	Your method	Measurements	Calculated density (g cm-3)
X	Report what you did. In particular, mention: • any problems you had and how you overcame them • whether you found two *different* methods for finding the density of the same substance.		

Task

Provide the dimensions of the room you are in, then calculate the mass of air in this room based on your determination of the density of air.

concept of air having mass and a density, a challenge is provided in the final section of this investigation for students to use their value for the density of air to calculate the mass of the air in the room in which they are working, which will also require them to calculate the volume of the room.

Other modifications that can be applied to science investigations to enhance student learning may include the following:

- Narrow the focus of the investigation to one of two possible outcomes—outcome A or B—before students commence the investigation. An example would be which paper towel adsorbs the most liquid: paper towel A or paper towel B.

- Get students to write their own description of the method after a discussion (so there is only one method).
- Get students to plan the method (as in Case Study 6.3).
- Provide instructions cut up in an envelope and ask students to arrange them in order. It is important that you discuss the aim first, or the task will be impossible.
- Include an 'unseen' question on the purpose or design of the investigation as part of the assessment. The students do this at the start (after reading through the investigation overnight) and hand it in on the spot.
- DON'T verbally point out things you want them to *read* (unless they need further explanation); instead, create the conditions that will encourage your students to get used to having to read.

THINKING TASKS TO PROMOTE STUDENT LEARNING IN SCIENCE INVESTIGATIONS

Below are a number of thinking tasks that can be used to further promote student learning in science when undertaking investigations.

Give students less information

Most teachers give students everything they need in order to do a science investigation. By withholding information or equipment, you can encourage students to make some judgements about whether they have everything they need. The following are some examples:

- Get students to design their own approach to an investigation. They will then need to make decisions about how to communicate this approach—for example, they could prepare a flow chart.
- If there are calculations that need to be made, get students to decide what needs to be calculated (e.g. do you have enough

material for the investigation) and how this may be done (e.g. prepare a flow chart or outline).

- Allow students to decide what data need to be recorded rather than telling them. Once they have recorded the data, they will also need to decide what needs to be done with the data and how to use them.

Reflect on aspects of the instructions

To facilitate students thinking about the investigation, you could get them to rephrase in their own words what the purpose of the instructions are and at the same time articulate why this instruction is important. This helps them to decide what instructions are important, what instructions need to be carried out in a particular sequence and what instructions can be carried out concurrently.

Additionally, if instructions are provided to students for an investigation, it may be worthwhile to get them to consider whether there are any instructions or any data that may not be necessary. When considering instructions, it is fundamentally important to get students to focus on the possible outcomes of what they are doing. Getting into the habit of making predictions prior to undertaking an investigation is one way to facilitate this.

Linking tasks

Provide opportunities for students to generate links with any concepts, ideas or information they already know. The use of diagrams such as concept maps, Venn diagrams and flow diagrams helps in such linking. Students should also be encouraged to link the investigation with concepts and skills from other learning areas.

For example, if they need to manage the data they have or make some calculations, encourage them to think about where they have done something similar before, such as in mathematics for calculations or other disciplines in terms of other ideas and concepts.

In addition, connecting the work they are doing in this investigation with work they have undertaken in other investigations also helps students to make valuable connections and recognise patterns that may be emerging—for example, that using a key in biology is similar to categorising substances in chemistry.

Reflect on the results

Good investigations also use common sense. It is worthwhile encouraging students to think about whether the results of the investigation seem reasonable, and whether they are in line with what was predicted or in line with scientific understandings of the concepts or ideas. Encourage students to think about what would happen if they altered one aspect of what they have done—how would this change the outcome? Using science to predict possible outcomes is an important part of scientific work that is often overlooked.

General processing tasks

Teachers can use a number of tasks to promote students' processing of information. These include:

- unseen questions at the start of a session
- constructing a diagram that highlights relationships such as concept maps, flow diagrams and Venn diagrams
- translating results into sketch graphs or pictures
- translating diagrams and graphs into prose or text
- practising sample calculations.

All these pedagogical approaches will assist students to become more focused on the processes and outcomes underpinning science investigations by linking these understandings with what they already know in ways that are meaningful to them.

CONCLUSION

It is important to remember why scientific investigations are important. This chapter articulates that there are some key features that are connected with the scientific process and support investigations being scientific in nature: empirical data, reasoning, avoidance of bias and peer review.

As Bybee (2006, p. 3) points out:

> Science is a human activity. Scientists may share some of the fundamental assumptions such as the essential place of data, the importance of logic, and the need to base their explanations on evidence, but scientists vary in their talents, imaginations, intuition, and courage.

As teachers of science, our task is to draw upon a range of investigation types and pedagogical approaches to make the various aspects of science come to life for our students.

DISCUSSION QUESTIONS

6.1 What role do creativity, imagination, trial and error, and mistakes and failure have in the investigative process? How do these constructs contribute to science learning?

6.2 What value does the inclusion of uncertainty as a feature of the investigative process have, and why is this important to our understanding of science?

6.3 How might the social and cultural embeddedness of knowledge be experienced and acknowledged through the investigative process?

6.4 Why do scepticism and peer review play such significant roles in the investigative process? How might you recreate and nurture these features as part of science learning in the classroom?

6.5 Which aspects of the investigative process and/or associated pedagogies do you feel you might need to learn more about as a teacher of science? How will you achieve this?

FURTHER READING

Concept of evidence website. Retrieved from http://community.dur.ac.uk/rosalyn.roberts/Evidence/cofev.htm.

Millar, R. (2002). 'Thinking about practical work'. In S. Amos & R. Boohan (eds), *Aspects of Teaching Secondary Science: Perspectives on Practice*. London: Routledge Falmer.

Rennie, L. (in press). Communicating certainty and uncertainty in science in out-of-school contexts. In D. Corrigan, C. Buntting, A. Fitzgerald & A. Jones (eds), *Values in Science Education: Reconsidering the Nature and Place of Values.*

REFERENCES

Bybee, R.W. (2006). 'Scientific inquiry and science teaching'. In L. Flick & N. Lederman (eds), *Scientific Inquiry and the Nature of Science*. Dordrecht: Springer, pp. 1–14.

Corrigan, D. (2015). 'Five levels of cognitive engagement'. Personal communication, March.

Gott, R., Duggan, S., Roberts, R. and Hussein, A. (n.d.). 'Research into understanding scientific evidence'. Retrieved from http://community.dur.ac.uk/rosalyn.roberts/Evidence/cofev.htm.

Harwood, W.S. (2004). 'An activity model for scientific inquiry'. *The Science Teacher*, 71(1), 44–6.

Royal Geographical Society (n.d.). *Fieldwork Techniques*. Retrieved from www.rgs.org/OurWork/Schools/Fieldwork+and+local+learning/Fieldwork+techniques/Fieldwork+techniques.htm.

Tripaydonis, T. (2008). 'Consumer science—is it real science?' In J. Loughran & A. Berry (eds), *Looking into Practice—Volume 3*. Melbourne: Monash University, pp. 85–7.

Vos, G. & Jones, R.M. (1986). 'The role of stolons and rhizomes in legume persistence'. In CSIRO Division of Tropical Crops and Pastures, *Annual Report*. Brisbane: CSIRO, pp. 70–1.

Wong, S.L. & Hodson, D. (2010). 'From the horse's mouth: What scientists say about scientific investigation and scientific knowledge'. *Science Education*, 93(1), 109–30.

Part III

Pedagogy and assessment

With a grounding in understanding the learner and what they bring to science learning as well as understanding science as it is practised and conceptualised in curriculum, an opportunity arises to further consider what this means for teaching. Any number of 'tips and tricks' resources are available to inform the activities you might bring to a science lesson, but we want to extend your thinking beyond this narrow focus. By thinking more deeply about what influences teacher decision-making and—more importantly—why, steps can be taken to not only improve your practices as a teacher of science, but ultimately enrich your students' learning of science.

Part 3 teases out two key areas for consideration:

1 *assessment*—recognising the different types and purposes of assessment and how they can be integrated into science teaching to enhance science learning
2 *pedagogical content knowledge (PCK)*—understanding the concept of PCK and leveraging from this to make more informed decisions about what matters in terms of your science teaching practices and why.

This section of the book is valuable because it requires you to shift your thinking from an activity-driven focus and concern about *what* you are teaching. Instead, you are encouraged to dig deeper into the craft aspects involved with science learning and teaching by questioning *why* you are doing what you are doing in the classroom.

CHAPTER 7

ASSESSMENT FOR STUDENT AND TEACHER LEARNING

DEBRA PANIZZON *and* **STEPHEN KEAST**

LEARNING OUTCOMES

After reading this chapter, you should be able to:

- identify assessment as central to our teaching, with a number of components aligning and linking together to ensure that students achieve intended learning outcomes
- recognise the different purposes of assessment
- draw upon a number of strategies that exemplify the ways in which we can monitor what our students understand and can do in science
- value the role of feedback for enhancing student achievement and self-efficacy
- demonstrate an understanding of how accountability affects and impacts teacher assessment in the classroom.

> In the classroom it is happening all the time. When the looks on the faces of the students, or their written work, or their oral answers to a question are appraised, the teacher is assessing. A quiz or written test is also an occasion for assessment, but it is only one among many possibilities. (Black, 2003, p. 6)

Think about the quote by Black and how closely it aligns with your views of assessment. What does Black mean by saying that assessment occurs all the time in our teaching? Surely assessment is just about getting students to complete tasks, activities, worksheets, investigations or tests that we, as teachers, then mark and hand back. Actually, **assessment** is so much more than this! In school, assessment in science refers to the process of gathering and interpreting information about the progress of students' learning in relation to scientific understanding, skills, values and attitudes (Hackling, 2004). As teachers, we need to monitor our students in an ongoing way using verbal, written and non-verbal cues to gain insight into what they understand, what is not quite 'clicking' and how our teaching might be altered to further support students in their learning. Hence assessment is not only about our students and their learning, but also about our learning as teachers and how we use the information we gain from students to plan our teaching.

assessment the process of monitoring student progress and learning

Views of assessment in school science vary considerably across primary and secondary teachers and students. For example, read the following excerpt from Timothy, a Year 11 student completing physics and a number of other subjects.

> *Interviewer:* What does assessment mean for you in science?
>
> *Timothy:* Well, at my school, it is pretty much used to

gauge whether you are at the top of the class or not. I guess it is used to let teachers know where kids are at but I don't think it is used for anything beyond that!

Interviewer: Do you see a link between assessment and learning?

Timothy: To the extent that it tells a kid whether they are getting the most out of it, yeah, but I think most students see it as giving you a deadline to learn stuff—so if you know there is a test in two weeks, you learn for that. In Years 11 and 12, we have testing for each topic we do. I think it has this focus because it is an easy way to assess. You ask kids questions and they get a score and you get ranked pretty much from that!

Interviewer: Is it driven by an end-of-year exam?

Timothy: Definitely. Everything is driven by the exam. The only thing we want is what is examinable. If something is not to be in the exam, we shut off! When you are in Year 11 and then Year 10 and then Year 9 everything is about the exam! But this is different in my other subjects, like English.

Interviewer: How is English different?

Timothy: Most other subjects I get more feedback and support from them. If you write an essay, they will tell you what you got wrong so you know how to improve next time. But in science, you have the test and the teacher will give the marks back and you have to really seek out more information. Assessment in science is really blocked into little pieces—topic with test done so you move on to the next one!

Timothy shares a great deal about what assessment looks like in science when compared with English, and ultimately how this affects his focus on learning in both subjects. For Timothy, if a concept or scientific idea is not in the examination, then there is little point in spending time learning it. If you think about your own experiences of teacher education, you have likely made similar kinds of judgements about assessment tasks based on the subliminal messages expressed by your lecturers, communicated in the unit readings and provided in the unit documentation. This is not at all surprising, given that the research indicates that *what is assessed* and *how it is assessed* provides the clearest indication to the learner about what is actually valued by a teacher or lecturer, and so where best to spend their energies to maximise achievement (Corrigan et al., 2013).

For some of you, these ideas may challenge your own ideas of assessment and your role and responsibilities as a teacher. For others, our ideas may sit quite comfortably. The prior conceptions you hold will ultimately affect how you engage with the chapter. To be clear, our key message is that assessment is central to our teaching—*what we teach, what the students experience as part of the teaching, and how we monitor their growth in understanding, skills, attitudes and values* must align and link together if we are to ensure that students meet the intended curriculum learning outcomes.

In this chapter, we begin by discussing the different purposes of assessment because this should be clear for us as teachers from the outset. Following this, we provide a range of strategies that exemplify the ways in which we can monitor *what our students understand* and *what they can do* in science. The key point in this section is that all strategies or tasks can be used in different ways to assess students, depending on the intention and purpose of the assessment as determined by the teacher. An important component of this discussion is the role of feedback for enhancing student achievement and self-efficacy. Finally, we discuss briefly the increasing accountability that teachers face around assessment and how this often impacts what happens in the classroom. This chapter sets out to challenge many of the preconceived ideas evident in some spheres of education about what assessment of students' scientific understanding looks like in the classroom and the types of tasks traditionally associated with the assessment of school science.

ASSESSMENT: DIFFERENT TYPES, DIFFERENT PURPOSES

To begin, let's explore the different types and purposes of assessment that are used when thinking about our teaching. As identified in Table 7.1, there are four main types, each with quite a unique purpose. These are likely to be very familiar.

While these types of assessment are mainstream in the literature, there are other ways to think about assessment. In particular, Paul Black and Dylan Wiliam (1998) from the United Kingdom initiated a major global shift in assessment in the late 1990s. Their ground-breaking work emphasised the highly integrative process of assessment in supporting students in their learning while informing teachers about their practice. This emphasis moved dramatically away from viewing assessment just for marks and the reporting of student achievement. Emerging from their work were the terms 'assessment for learning' (equivalent to formative assessment), 'assessment of learning' (summative assessment) and **assessment as learning**. This last term is relatively new and quite powerful. It refers to assessment as a means of encouraging students to self-reflect on their own learning, and to become more **metacognitive** and responsible for their future learning as they monitor the strengths and gaps in their own learning (Black & Wiliam, 1998). So, rather than assessment being done *to* students, the students use assessment to *guide their own learning*.

assessment as learning puts the students in the centre of their own progress in terms of learning

metacognitive the ability of students to reflect on their own thinking and learning processes

Table 7.1 Main types and purposes of assessment

Types	Definition	Purposes
Summative	Summative assessment measures student achievement as grades that can be reported.	The teacher collects evidence to assess student learning against achievement standards or outcomes so that student achievement can be reported against some kind of norm (Millar, 2013). Usually tasks are completed or submitted at designated times for marking, with the results collated and used to produce a final grade that is communicated to the student and parents. This is the type of assessment with which we are most familiar from our own schooling and university experiences.
Formative	Formative assessment is the ongoing monitoring of student progress, and does not result in grades.	The teacher informally apprises each student about their learning progress in an ongoing manner by identifying those areas or aspects that are understood, along with possible gaps in learning where the student might need to direct their energies towards improving their understanding. Teachers may read responses but no marks are collected (Millar, 2013).
Diagnostic	Diagnostic assessment gauges student knowledge prior to a teaching sequence.	The teacher attempts to gauge students' conceptions and understandings *prior* to teaching a particular topic or unit of work so that teaching can build upon what is known and understood by students (Matters, 2006). Usually teachers undertake this type of assessment informally in their classrooms through questioning and brainstorming. While the students may be provided with some feedback, this type of assessment is really about informing the teacher, who can then plan appropriate learning experiences and opportunities to align with the understandings of individual students in a more targeted manner.

Types	Definition	Purposes
Evaluative	Evaluative assessment occurs on a broader scale, comparing student achievement over time.	The teachers' and students' learning is considered more generally as educational systems attempt to gain a sense of the overall achievement trends of cohorts of students over time. Examples of these forms of assessment in Australia include the National Assessment Program Literacy and Numeracy (NAPLAN), Trends in International Mathematics and Science Study (TIMSS) and the Programme for International Student Assessment (PISA). The purpose of these types of assessments is to monitor the performance of schools and educational systems against specified standards while benchmarking against international trends.

ASSESSMENT STRATEGIES FOR THE SCIENCE CLASSROOM

So what kinds of tasks and strategies can we use to actually assess our students and how do we decide which ones are most appropriate? In this section, we provide a range of possible assessment strategies, explaining how each might be used for different purposes. For each of the strategies, we provide specific examples along with ways in which they might be implemented in your teaching.

Brainstorming

Brainstorming is the process of generating a range of ideas or solutions, usually through open group discussion. Brainstorming is often done informally at the beginning of a lesson or topic to determine the prior knowledge of students. There are many ways to conduct brainstorming, from asking students for ideas, and writing or recording these on the interactive whiteboard, to giving students some individual time to think about their own ideas before contributing in class.

The idea of brainstorming is to draw out information and knowledge held by your students for a particular science concept or topic.

Capturing the ideas that emerge from brainstorming can be done in many ways. For example, you might summarise ideas on an interactive whiteboard with the summary captured to refer to later. Alternatively, you might ask students to write questions they would like answered about a science topic before it is actually taught. Such questions can be posted onto a 'Question Wall' or 'Word Wall' and maintained while the topic is being taught. When a student thinks a question has been answered, they go to the wall, select the question and provide the answer for it. In this way, students are encouraged to be metacognitive (assessment as learning) and responsible for their own learning, and to take an active part in recognising when they have learnt something new (formative assessment). If appropriate, students might be encouraged to participate in these activities by allocating marks for class participation when a question on the Question Wall is answered (summative assessment).

Think–pair–share

One of the major problems with brainstorming is that the same students contribute all the time, creating an environment where some students may feel that they have no ownership of the ideas or questions being raised. This challenge can be overcome by employing a 'think–pair–share' strategy. Think–pair–share is a collaborative learning activity used to stimulate thinking around a particular idea. During the 'think–pair–share', students write or draw their ideas individually. During the 'pair' phase, students pair up with another student or even in small groups around the table, and build a shared list of terms or questions. Often, in this phase, new ideas emerge prompted from the discussion between the 'pair'. In the 'share' phase, the pair or group ideas are shared with the entire class and usually recorded on the interactive whiteboard.

Visual representations

These kinds of representations can be used with students on their own or as a way of getting students to work with some of the ideas generated as part of an initial brainstorming session.

Concept maps

Concept maps are visual displays identifying ideas or concepts and the links existing between them. These are displays of concepts for an area of study drawn as shapes on a page with clear links identified between each concept. Each pair of links has arrows showing the direction of the connection along with a comment that describes how the concepts are connected. A student's understanding of a topic or idea is demonstrated by the number of concepts identified, along with the simplicity or complexity of the links and descriptions between these concepts (i.e. see the differences between Figures 7.1 and 7.2). These maps explore the concept of particle movement and the impact of temperature on this movement. In looking at

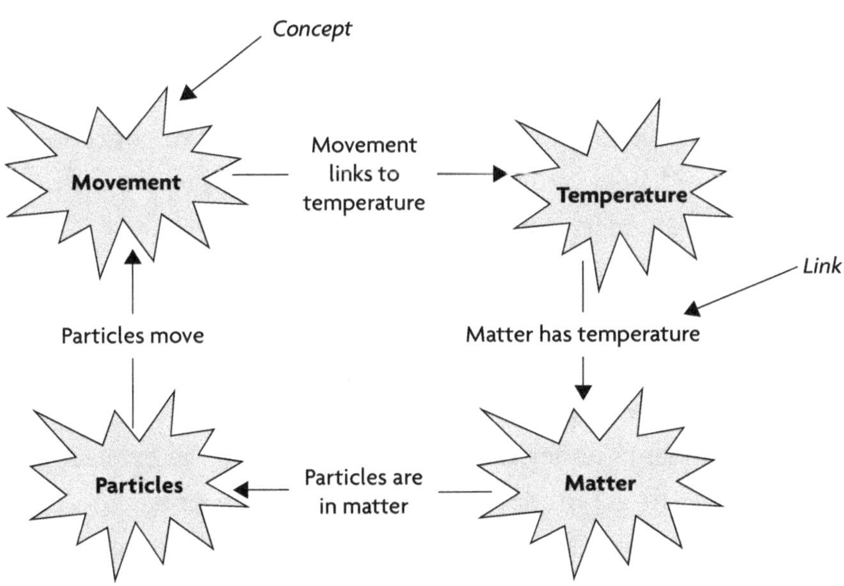

Figure 7.1 Simple concept map

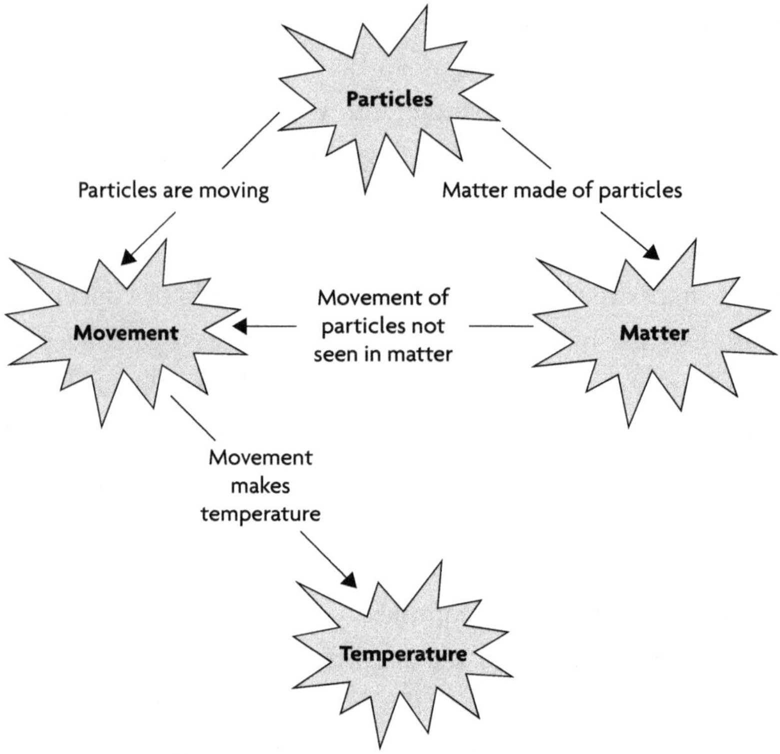

Figure 7.2 Concept map of increased complexity

these two maps, a greater level of understanding is demonstrated in Figure 7.2. Concept maps often reveal the alternative conceptions (see Chapter 2) held by students in science, so are useful for both primary and secondary students.

Concept maps might be applied in a variety of ways to assess students. For example, they might be used at the beginning of a topic to determine students' prior knowledge (diagnostic assessment). Alternatively, they might be incorporated at the end of a topic to determine students' understandings after teaching. An extension of this might be to get students to construct a concept map at the beginning of the unit and then be given the same map again at the end of the unit to alter (using another colour) to show changes in their understandings and thinking. In this way, the concept map is used to demonstrate growth in understanding that can be quite powerful for the student (assessment as learning).

Venn diagrams

These are also visual representations of intersecting shapes. A Venn diagram is a visual way of showing the similarities and differences between different concepts or ideas.

The intersections represent common characteristics of two (or more) scientific concepts or ideas. Figure 7.3 provides an example of a simple diagram that considers the similarities and differences between mammals and fish.

These diagrams are a great way for students to compare and contrast key concepts or ideas in science. While this figure shows two circles, with some practice students will be ready for three circles—for example, mammals, fish and reptiles.

Interactive whiteboards provide an interesting dimension for working with students. Teachers can collect relevant scientific information from a brainstorming session on the board, then ask students to suggest categories or concepts, and draw either a concept map or a Venn diagram, depending on which is most appropriate. Terms,

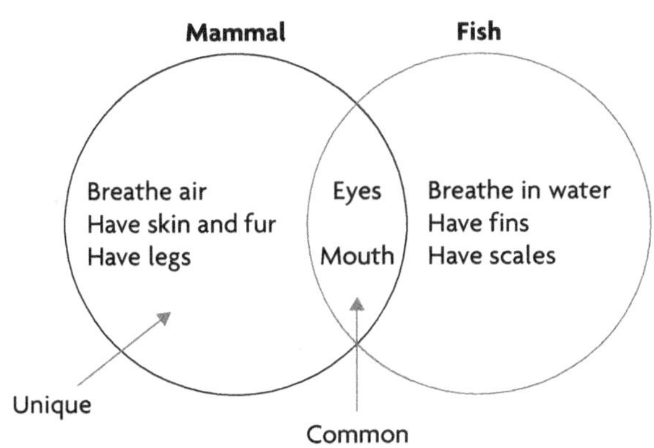

Figure 7.3 Venn diagram

links, circles and descriptors can all be dragged easily as students explain the reasoning behind their thinking.

Portfolios

A portfolio is a vessel that holds different artefacts—the vessel can be physical or virtual (electronic). Portfolios can be used in several ways that acknowledge students' ongoing learning. With careful implementation, they help to encourage students to become more metacognitive (assessment as learning), taking greater responsibility for and having awareness of their own learning journeys. Usually, a portfolio consists of a display folder or electronic folder where students can collect or submit samples of their work. This has several advantages over other types of assessment because it allows both teachers and students to view the learning journey and how it has altered over time. Often students can get to the end of a period of time and reflect that they have not learnt very much. However, a portfolio provides a record of evidence of their learning. Portfolios are particularly useful when they include before and after items (e.g. concept maps or Venn diagrams) that clearly demonstrate changes in students' thinking over time.

As an assessment task, portfolios can be used as formative or summative assessment, depending on how the teacher sets this up with the students. For example, the teacher might allow students to place their best work in their portfolio to demonstrate what they can do. The types of work that might represent their level of engagement and learning could be a practical activity or representation, a set of questions or story. In this way, the teacher shares intellectual control with their students by allowing students to select the pieces placed in the portfolio. Of course, the substance of these portfolios will vary depending upon what the teacher and student negotiate as being the focus.

Learning logs

A learning log is a personalised resource produced by students to record their responses to learning challenges or activities along with reflections about their learning journeys.

These are guided reflection tools that encourage students to record and recognise not only their learning in science but also the strategies that either assist or hinder their learning. They are usually designed around specific reflective questions to which students respond over the course of a teaching period. When first used, students often focus on what they did in class, like a diary entry, rather than what they learnt, how they learnt it and what affected their learning. So it is important for us to scaffold their entries by asking students to respond to a few questions, such as:

- What did I learn today?
- What helped me learn?
- What hindered my learning?
- What am I confused about?
- What don't I understand?

This type of task supports all students in becoming more meta-cognitive and aware of their own learning—what works for them and what does not. Used together with portfolios, learning logs are a powerful way to assess students that moves away from the traditional focus on knowledge and skills to concentrating on the learning process. While the portfolio represents a record of what a student can do, a learning log is a student's record of how they learnt, what they have learnt and which aspects really impacted their learning along the journey. Initially, learning logs are best used for formative assessment but as the students gain experience and confidence in this type of writing, they can also be used for summative assessment.

Predict, observe, explain

Predict, observe, explain (POE) is a learning strategy that stimulates students' thinking about scientific phenomena. Proposed by White and Gunstone (1992), the POE strategy is widely used in the teaching of science regardless of the year level of the student. A POE has three separate stages:

1. *predict*—what do you think will happen?
2. *observe*—what happened?
3. *explain*—how did it happen?

1. Students are introduced to a situation where a change will occur as the result of some action—for example, a small car held by you on top of a ramp. This activity might be used to introduce a topic on forces with younger students, allowing them to explore the range of pushes and pulls that are exerted on the car. The students are then asked to make a *prediction* about what might occur once your hand is released. For younger students, this could be a drawing while older students can write their responses.
2. The change is then made—the car is let go to move freely. Students then *observe* what happened. It is valuable to get the students to record their own observations too, as they often notice surprisingly different things with even the simplest of practical experiences. You may want to do this phase of the activity a number of times so that students get multiple chances to make their observations.
3. Students are now ready to propose an *explanation* for their observation. An important part of this phase is encouraging students to explain any differences between their initial prediction and their actual observation.

POEs can be done with students from the primary school years through to Year 12, with increasing complexity. They provide an engaging way of introducing science concepts while also nurturing the development of scientific skills. Expecting students to make a prediction before they observe the change creates a greater focus of attention on the task at hand. Students being able to then share their ideas and explain the links or disconnect between their predictions and observations paves the way for greater engagement in learning the science behind their experience and why this occurred. Given their nature, POEs are probably more aligned to formative assessment.

Practical work—inside, outside and beyond the school

Practical work is a broad term that refers to all types of activities that might be undertaken within a laboratory. It also includes activities that can occur in the school playground, in the field or at educational sites, such as museums or the zoo. Student engagement in practical work, regardless of its location, is a key to learning in science. Practical work in science has many purposes, including teaching through inquiry, skill development (e.g. how to collect appropriate data), a real experience of a **concept** (e.g. observing chambers of a heart), exploration (e.g. what happens when . . .), confirming a theory (e.g. dropping objects from various heights) or demonstrating a scientific concept (e.g. electricity requires a complete circuit). Scientific concepts can be basic (e.g. plant) or more complex (e.g. energy), or made up of many basic concepts linked together. All concepts are constructed and owned by the learner, which is what underpins the constructivism in learning science. The research, however, indicates that 'hands on' does not mean 'minds on' or that students are cognitively engaged in the experience, so it is critical for the teacher to consider the purpose of the practical activity because this ultimately will determine how it might be assessed.

concept a mental abstraction that represents patterns or relationships among objects, events and other concepts

Within the primary classroom, there appears to be a great deal more flexibility in the ways that primary teachers go about assessing their students in relation to practical work. Use of representations, drawings with explanations or concrete artefacts (e.g. a diorama of the solar system) are commonly used to assess younger students. In contrast, the assessment of practical work often becomes problematic for many secondary science teachers, who tend to hold fairly traditional approaches about how this is best achieved. The practical report, with its very rigid structure, is probably used most frequently. Yet this method of monitoring student progress may not be the most appropriate. For example, assessing students' microscope skills is best achieved by watching them operate the microscopes, viewing their specimens and observing their drawings. Why expect the students to write a full laboratory report? Secondary students do need to be able to write a report, but how many of them do they need to complete? These types of rigid formats based upon recipe-style practicals promote science that limits students' thinking while results are readily available on the internet for submission as their own work.

With a little ingenuity, there are several ways of changing the usual practical experience to ensure that students are thinking from the outset and are cognitively engaged, and that student achievement is being appropriately assessed. These include the following.

- *Inclusion of an 'unjumbling task'*, where the steps for the method of the practical are jumbled up in an envelope. Before the students can actually start the practical, they must first work out the correct order of the steps. This requires the students to process the information provided, then sort and order it logically. This can be very insightful if several steps can be done in any order because students then need to explain the reasons for undertaking the steps in their preferred order. Higher marks can be allocated to this section or it might be that only this section is assessed summatively to determine the degree to which students understand the process required for the practical.

- *Inclusion of two or three challenging questions at the end of a practical* that require students to apply and extend what they have learnt from the practical. These might relate more to the real-world application of the underlying concepts being explored during the practical. Again, these responses might be allocated more marks than other sections of the report, such as the method that is routine in its nature.

The key here is to provide variety so that practical work is not routine and students are never sure what kinds of additional information and variables need to be considered in preparing for the activity.

Slowmation

Slowmation is a narrated animation used to explain a concept or tell a story. Slowmation is a tool that is unique and different from other assessment tasks or strategies. It provides deep insight into the ways in which your students conceive of scientific concepts or ideas. In brief, Slowmation involves the students taking digital still photos and combining them into a movie to make an animation. The value of this method of assessment is that students have to:

- consider their understanding of the whole process and how they will represent it
- break it down into separate components and recognise how they understand each one
- rebuild the ideas into a coherent whole to be made into a movie that can be understood by their peers.

Initially, students complete a storyboard or chunking sheet, where they map out the major scientific concepts or ideas on paper and then conceptualise their movie production. Once completed, the chunking sheet gives the teacher their first insight into how students understand the process. Feedback at this stage, before the production of a model, is important, but it should be provided as questions rather than answers—for example, 'What do you think happens between this stage and the next stage?' or 'How do you think it happens?'

Given this feedback, students modify their chunking sheets and begin model production and photo taking, which requires the equivalent of two lessons (100 minutes). As the models are built and modified, photos are taken while the teacher monitors progress and asks questions to maintain focus on the science. Once the photos are taken, students begin assembling their movies with MovieMaker or iMovie. They can add sound, labels, titles and narration at this stage. Afterwards, students show their movies to the rest of the class, which provides a critical opportunity to determine students' understandings, as careful questioning at this point can reveal alternative conceptions. These movies have been found to be highly revealing regarding students' understandings of scientific concepts (Kidman, Keast & Cooper, 2012).

Given the time it takes for students to prepare, make and show their Slowmation movies, this task is best used for summative assessment. However, teachers using Slowmation can allocate marks for particular sections, such as marking the chunking sheet before the model building begins, which then helps to focus student attention on the accuracy of their science.

Tests and examinations

Tests and examinations refer to specifically designed assessment tasks that comprise a number of items that assess the degree to which students have understood aspects of their learning. While used for summative assessment, they are equally appropriate to be used formatively and diagnostically with students.

Tests are used commonly by secondary science teachers and rarely by primary teachers as a means of summative assessment. Yet they can be used really effectively for diagnostic or formative assessment in ways that enhance student learning. A short impromptu quiz (test) might be given at the beginning of a teaching sequence to determine the students' prior knowledge along with gaps in understanding and prevailing alternative conceptions (diagnostic). Alternatively, a quiz can be used at the beginning of a lesson to determine what the students have understood or what they are confused about from the previous lesson. The information gained from this quiz provides the student with some idea about what they know while informing the teacher about where best to direct teaching. However, most tests in science are used at the end of a teaching sequence or unit of work to gauge and report on student achievement.

It is our view that tests are useful forms of assessment—it is actually the quality of the items that makes a test 'poor'. Before preparing a test, the teacher must think carefully about its purpose and what the students are expected to demonstrate in the test. Is it about content? Is it about requiring students to apply their scientific understanding? Is it about capturing their skills and ability to think about the processes of science? Answers to these questions will help improve the quality of the items included by the teacher. The production of equitable and sound quality tests takes considerable time and practice.

Strategies are available to raise the level of information that tests provide around students' understandings. With small changes and some careful thought, it is possible to change mediocre questions into items that are so much more informative for both the student and the teacher. As an example, multiple-choice questions are commonly used by science teachers because they are easy to mark and you can cover a lot of scientific content. However, they are limited in what they actually demonstrate about student understanding. A study conducted by the National Science Foundation Project 2061 (AAAS, 2007) found that many students guessed the correct answer by simply using cues in the question itself. To counteract this possibility, a follow-up question might be included, such as 'Explain why you selected this particular answer'. This then forces students to justify their response. Careful allocation of marks, such as one mark for the actual correct response but two marks for the explanation, ensures that students who actually understand the science are rewarded appropriately with higher marks.

The bottom line is that tests—like all assessment tasks—should offer students the opportunity to demonstrate their understanding of the concepts, processes and skills they have been taught. Some teachers use tests to introduce new ideas that may not have been addressed adequately in class and so deliberately target what the students *do not know*. This practice is not equitable, and is unfair to the students.

In overview, we consider that any task can be used in many different ways to assess students' understandings in science. While most teachers can find tasks easily, a fundamental building block often overlooked is that exploring depth in understanding requires insightful and thought-provoking questions. In other words, we need to use questions that elicit the degree of scientific understanding held by our students. This seems so obvious and yet, in the day-to-day routine of the classroom, it becomes easy to fall back on simple closed questions that are useful for some purposes but not for actually challenging or extending the thinking of our students (Panizzon & Pegg, 2007).

Clarifying expectations and providing quality feedback

Given that assessment is an integral part of the teaching and learning process, it is only fair that the expectations are made explicit for students. This is often achieved by identifying the:

- aim of the task
- learning outcomes (including scientific skills and processes) to be assessed
- criteria to be used in assessing students' progress.

Interestingly, some teachers are uncomfortable about making these criteria explicit for students, holding the view that, 'If students know what we are marking, what stops them all from getting 100 per cent?' The reality is that even if students are provided with these criteria, not all students will be able to use this information to achieve a high result. But as teachers, we must ensure that equity prevails at all times in our classrooms.

Rubrics are a common way of communicating the criteria to students in ways that allow teachers to align feedback against each criterion. Broadly defined, feedback refers to the information provided by a teacher, peer, parent or experience that helps to reduce the discrepancy between what a student understands (where the student is at) and what the student actually needs to understand (where the student should be). However, there is quite an art to providing detailed feedback that informs students and enhances their achievement in the longer term. Hattie and Timperley (2007) found that teachers who did provide detailed feedback answered the following four questions for students on a regular basis:

- What can the student do?
- What can't the student do?
- How does the student's work compare with that of others?
- How can the student do better?

It was of particular interest that highly effective teachers focused on the last question. Its value is that it gives students clear direction and guidance about *how to improve their own learning*. But this is also the hardest question for teachers to address because it requires knowing your students so that you are able to give each student specific and relevant information.

Aligning assessment with quality learning and teaching

Having considered different strategies for assessing our students, how do we choose which one(s) to use? To answer this question, we need to ensure that *what we teach* (curriculum), *how we teach it* (our pedagogies and practice) and *what our students have understood* (assessment) are inextricably linked (Fensham & Rennie, 2013). It is only when we think about what we expect our students to understand and demonstrate in terms of skills and scientific processes that we can begin to think about how best to teach in ways that support students to build and develop these components. Hence assessment is pivotal in planning our teaching, and should not merely be an afterthought at the completion of a teaching sequence. Consider the points made by experienced secondary teacher Caroline based upon her own experience:

> Finding tasks or activities to assess students is easy—simply use Google! But does this task align with your teaching goals and what you want the students to know and do? Often we teach, give students learning opportunities, then assess them on something different or only on part of what we have taught them. So where is the equity of this type of assessment for our students?
>
> When I first started teaching, it was really just about finding out what the students knew. They were either correct or not! But over time I thought more about what I was doing as part of my teaching and the experiences I offered—how was I enhancing student learning? What can I change and do better?

As discussed in this chapter, assessment is very complex because it means different things to different people. Adding to this complexity, in the last decade assessment has moved away from just being about teachers and their own students as increasing levels of accountability have become a focus for educational authorities and the government. With the implementation of greater national and international testing (e.g. NAPLAN, PISA, TIMSS), teachers are often conflicted between best practices to support student learning in their own classroom while trying to help students attain high results in these externally designed and marked tests (Black, 2013; Cowie, 2009).

Evidence is available about the impact of high-stakes assessment in countries like Australia, the United Kingdom and the United States, where there is a culture of *teaching-to-the-test* (Linn, 2000; Millar, 2013). In Australia, this growing culture is so dominant that teachers have been reprimanded, and even dismissed, for helping students during the NAPLAN tests and even altering student results. This demonstrates the increasing pressure being placed on teachers to attain high results for their students on these large-scale, highly accountable tests.

CONCLUSION

In this chapter, we have deliberately set out to challenge the notion that assessment occurs at the end of a teaching sequence and is solely for measuring and recording students' marks for reporting purposes. As demonstrated here, we view assessment as a pivotal component of informing students in an ongoing way about what they know and can do while guiding teachers about where best to direct teaching to support students in restructuring and building their existing scientific understandings. Hence assessment should be occurring all the time in our classrooms.

Critically, assessment in science can be achieved using a range of tasks or strategies, and should not be limited to the more traditional methods—namely worksheets, tests, examinations and scientific reports. In closing, we consider that any task can be used for summative, formative or diagnostic assessment—the bottom line is that, regardless of the task, quality assessment informs both students and teachers. As a teacher, it is your role and responsibility to select appropriate tasks that will help all students demonstrate their learning in fair and equitable ways. These points are reinforced in the following quote:

> Quality assessment requires clear and valued goals for student learning as the assessment targets quality tools for gathering evidence of student learning, sound interpretations of the evidence, and quality uses of the information to guide instruction and provide students with useful feedback. (Gearhart et al., 2006, p. 5)

DISCUSSION QUESTIONS

7.1 In what ways has this chapter impacted your thinking about the purpose of assessment?

7.2 The role of teachers in the development and use of assessment is well understood, but how might students use assessment as a learning tool?

7.3 Following on from the previous question, what role could students play in the design of assessment and/or the assessment process?

7.4 How might you summarise the features of quality assessment?

7.5 Based on your own experiences of assessment, both as a student and a teacher, what might be some factors you will consider when next assessing science learning?

FURTHER READING

Corrigan, D., Gunstone, R. & Jones, A. (eds) (2013). *Valuing Assessment in Science Education: Pedagogy, Curriculum, Policy*. Dordrecht: Springer.

Heng, J. (2015). *Learning to Teach with Assessment: A Student Teaching Experience in China*. Dordrecht: Springer.

Wiliam, D. & Leahy, S. (2015). *Embedding Formative Assessment: Practical Techniques for K–12 Classrooms*. West Palm Beach, FL: Learning Sciences International.

REFERENCES

American Association for the Advancement of Science (AAAS) (2007). 'Science literacy for a changing future'. *2061 Today Newsletter*, 17(1), 1–8. Retrieved from www.project2061.org/publications/newsletters/2061Today/archive.htm.

Black, P. (2003). 'The importance of everyday assessment'. In M.J. Atkin & J.E. Coffey (eds), *Everyday Assessment in the Science Classroom*. Arlington, VA: NSTA Press, pp. 1–12.

—— (2013). 'Pedagogy in theory and in practice: Formative and summative assessments in classrooms and in systems'. In D. Corrigan, R. Gunstone & A. Jones (eds), *Valuing Assessment in Science Education: Pedagogy, Curriculum, Policy*. Dordrecht: Springer, pp. 297–329.

Black, P. & Wiliam, D. (1998). 'Assessment and classroom learning'. *Assessment in Education*, 5(1), 7–74.

Corrigan, D., Buntting, C., Jones, A. & Gunstone, R. (2013). 'Valuing assessment in science education: An introductory framework'. In D. Corrigan, R. Gunstone & A. Jones (eds), *Valuing Assessment in Science Education: Pedagogy, Curriculum, Policy*. Dordrecht: Springer, pp. 1–9.

Cowie, B. (2009). 'The evolution of assessment purposes and practices'. In S. Richie (ed.), *The World of Science Education: Handbook of Research in Australasia*. Rotterdam: Sense, pp. 235–48.

Fensham, P.J. & Rennie, L.J. (2013). 'Towards an authentically assessed science curriculum'. In D. Corrigan, R. Gunstone & A. Jones (eds), *Valuing Assessment in Science Education: Pedagogy, Curriculum, Policy*. Dordrecht: Springer, pp. 69–100.

Gearhart, M., Nagashima, S., Pfotenhauer, J., Clark, S., Schwab, C., Vendlinski, T., Osmundson, E., Herman, J. & Bernbaum, D.J. (2006). *Developing Expertise with Classroom Assessment in K-12 Science: Learning to Interpret Student Work—Interim Findings from a 2-year Study*. CSE Technical Report 704. Retrieved from www.cse.ucla.edu/products/reports/R704.pdf.

Hackling, M.W. (2004). 'Assessment in science'. In G. Venville & V. Dawson (eds), *The Art of Teaching Science*. Sydney: Allen & Unwin, pp. 126–44.

Hattie, J. & Timperley, H. (2007). 'The power of feedback'. *Review of Educational Research*, 77(1), pp. 81–112.

Kidman, G., Keast, S. & Cooper, R. (2012). 'Responding to the 5Rs: An alternative perspective of slowmation'. *Teaching Science*, 2(58), 24–30.

Linn, R.L. (2000). 'Assessments and accountability'. *Educational Researcher*, 29(2), 4–16.

Matters, G. (2006). 'Using data to support learning in schools: Students, teachers, systems'. *Australian Education Review*, 49. Retrieved from http://research.acer.edu.au/cgi/viewcontent.cgi?article=1032&context=resdev.

Millar, R. (2013). 'Improving science education: Why assessment matters'. In D. Corrigan, R. Gunstone & A. Jones (eds), *Valuing Assessment in Science Education: Pedagogy, Curriculum, Policy*. Dordrecht: Springer, pp. 55–68.

Panizzon, D. & Pegg, J. (2007). 'Assessment practices: Empowering mathematics and science in rural secondary schools to enhance student learning'. *International Journal of Science and Mathematics Education*, 6, 417–36.

White, R. & Gunstone, R. (1992). *Probing Understanding*. London and New York: The Falmer Press.

CHAPTER 8

PEDAGOGICAL CONTENT KNOWLEDGE

JOHN LOUGHRAN

LEARNING OUTCOMES
After reading this chapter, you should be able to:

- understand the concept of pedagogical content knowledge (PCK)
- develop ways in which you develop and then communicate your own PCK
- consider the many opportunities PCK offers for your professional learning.

DEVELOPING SCIENCE TEACHERS' PROFESSIONAL KNOWLEDGE

Pedagogical content knowledge (PCK) is an important aspect of teachers' professional knowledge of practice. In many ways, knowledge is often misinterpreted in relation to teaching, and can sometimes be seen as less important (or useful) than the ability to 'do teaching'. The theory–practice divide is often invoked when considering teachers' knowledge of practice, yet as a profession, there is little doubt that just being able to 'do teaching' is not enough to signify genuine expertise; rather, there is great value in being informed about the underpinnings of, and the reasons for, the 'doing'. The 'what', 'how' and 'why' of practice all matter, and combine in the overt expression of knowledge *of*, *in* and *for* teaching.

Although teachers inevitably spend a considerable period of time, energy and effort developing their armoury of teaching procedures, approaches and strategies, these alone should not be seen as the total package in relation to their professional knowledge. It is important that the pedagogical reasoning, the understandings of students' learning, and the knowledge of context and content are understood as central to shaping what teaching looks like and how it impacts students' learning. Therefore, teachers' professional knowledge of practice is not just about 'doing teaching'; it is about the what, how and why that shape the product 'seen' as teaching. That means that theory and practice are not in competition, but rather are complementary, and that together they offer windows into teachers' expertise. PCK is a specific example of how teachers' professional knowledge of practice extends beyond the general and applies to the specific and particular knowledge, skills and abilities essential to the development of teaching and learning within a subject area in its own right. In this chapter, we develop the idea of PCK and how it may be captured and communicated to other professional teachers.

Teachers' professional knowledge of practice needs to be recognised, rewarded and highly prized. In that way, arguments about what it means to be a teacher and how expertise might be developed must go well beyond approaches to 'up-skilling' or implementing mandated curriculum and/or policy changes. Teachers' professional knowledge of practice is about what matters with regard to knowing what to do, how to do it and, crucially, why it matters. We begin this chapter by entering a science teacher's classroom through a case study in order to set the scene for teacher thinking about teaching how the moon revolves around the Earth.

CASE STUDY 8.1: Learning to make the abstract concrete

One session that is vividly etched in my mind was the time I decided to use a role-play that I hadn't used before. I had noticed that, in the past, most of the role-plays only involved a small number of students and so I was conscious of the need to have more participants involved in *doing* things—to have less observing with no specified task to complete.

At the time, I was involved in a project that was documenting exemplary science teaching, so I decided to use a role-play from that project, which I knew would involve the whole class. On paper, it was a good role-play.

The role-play was designed to explore the way the Moon revolves around the Earth. I started off in the normal way by getting students' prior views, listing them on the board, pushing and probing their ideas and generally trying to create a sense of interest in the topic. The next phase was to set up the problem. It was something like, 'The Moon rotates on its axis once during its revolution of the Earth, so how is it that we only ever see the same face?'

I expertly stumbled through a teaching procedure (that I usually did not like) in a content area (with which I was unfamiliar) [when] it finally dawned on me that we had to go outside if we were to have enough space to do the role-play; we would never manage the role-play in the confines of the classroom. When we spread out [outside], I quickly organised everybody into their positions. I knew it couldn't be too hard. All I needed was a small group (four to five students) to make a circle by linking their arms while standing back to back—so that they would all be looking outwards—that group would be the Moon. Then the same formation was needed for the remainder of the students and they would be the Earth. So far, so good.

'Okay, Earth, start . . .' The role-play had begun.

'Now Moon, start a revolution around the Earth, but while you do it, you have to organise yourselves to do only one rotation on your own axis. So Janet, from where you are now, you need to be facing the Earth again when you get back to that spot. Put a mark on the ground so you know where that will be. Okay, off you go,' I said with great confidence.

I stood back and watched the role-play unfold. As the Moon slowly rotated in its revolution of the Earth, I tried to visualise how it could be that if it rotated only once on its axis that only one side of the Moon would be seen from the Earth. I couldn't do it, though. I just could not manipulate that in my mind to understand how it could be possible. But as I stood there watching it, I could see it happening before me. The abstract became very concrete.

When the Moon arrived back at its starting point, I asked how the role-play had worked—I was now moving away from the prescribed script; my curiosity had been piqued. Liz (a part of the Earth) and Athena (a part of the Moon) said they didn't get it. Having just watched it unfold before me, I decided that they would 'get it' if they could see it happening, so I got them to stand out and watch while we repeated the role-play and I took Athena's place in the Moon.

As the role-play unfolded again, I actually felt what it was like to be the 'dark side' of the Moon, and although in my mind I still couldn't really visualize what was happening, being a part of the action made it very real for me. Suddenly I got what it meant to be involved in a role-play [and see how it could make abstract ideas concrete]. Suddenly I saw a number of important pedagogical insights. Suddenly content matter started to take new shape as a developing understanding slowly emerged. Suddenly, our class became alive with learning and I was part of it.

After the class, I mused about the episode again—what I knew (or thought I knew) before the experience was dramatically different from what I knew after the experience. Being involved in the experience was different from directing it for others. Abstracting the learning from this experience to other situations was intellectually challenging and engaging. What I previously knew, I now understood. (Loughran, 2006, pp. 24–6)

Case Study 8.1 illustrates how developing knowledge about practice can occur and, if attended to, how that process can actively promote powerful professional learning and the development of a teacher's knowledge of practice. This vignette suggests that there is a major difference between using a role-play as an 'activity that works' and using it because the role-play helps to enable the particular content to make sense to students; making the abstract concrete in this case is recognition of an issue with learning about the specific content.

As Case Study 8.1 suggests, different teachers could interpret and respond to the situation in different ways. One response could be by using the role-play as an activity that works because the students had fun and enjoyed doing something different. If that was the case, then the role-play is simply a fun activity. However, as Case Study 8.1 makes clear, when a concentration on the issues associated with learning the science is the focus, then the role-play takes on new meaning as the teacher begins to see beyond the teaching procedure as an activity and into the reasons why it works the way it does—particularly in terms of the nature of learning.

THINKING ABOUT SCIENCE LEARNING AND TEACHING

This book highlights science education research that explores a number of interesting features of learning that are crucial to understanding quality in science teaching, such as an understanding of the nature of science (NoS), the consideration of context and relevance in which to embed learning, creating conditions for purposeful learning and a broad and purposeful approach to assessment. Another example, discussed in Chapter 2, is the long history of students' misconceptions/alternative conceptions (Wandersee, Mintzes & Novak, 1994), which has highlighted that although students' alternative conceptions might well be brought to the surface in a science class, being able to change alternative conceptions can be exceptionally difficult. One way of explaining why students' alternative conceptions are difficult to change relates to the fact that (to the student) they appear to explain everyday phenomena. Therefore, as a consequence of their apparent explanatory power, alternative conceptions tend to be reinforced rather than refuted, and so can become quite robust—and robust ideas often have strong 'staying power'.

Posner and colleagues (1982) offer an explanation about what it might take to challenge, and therefore shift, an alternative conception. Their conceptual change model is based on a view that the first step in conceptual change is driven by a sense of dissatisfaction with an existing conception—the idea that the conception no longer satisfactorily explains a phenomenon. With a sense of dissatisfaction with an existing conception, an openness to actively pursue an alternative explanation may well emerge (without dissatisfaction, it is less likely that a student would feel a need to change their thinking). Thus dissatisfaction can be seen as a form of motivation for change.

A common example of this process is thinking about the concept of particles. There are many students aged from seven to eleven who have no problem with water disappearing when it is poured into sand. This is because they know that if they dig deep enough at the beach, the water will come back. It disappears, but it can come back. The challenge here is to get them to think about where the water went and how it can be retrieved. Such an explanation relies on students having an understanding that substances are made up of particles and it is important to understand how those particles are packed. A common way to address this need to explain this phenomenon of disappearing water using particles is to use a predict, observe, explain (POE), where there are three test tubes made up in the following way:

- test tube 1: 50 ml water + 50 ml water = ? ml water
- test tube 2: 50 ml methylated spirits + 50 ml methylated spirits = ? ml methylated spirits
- test tube 3: 50 ml water + 50 ml methylated spirits = ? ml water and methylated spirits mixture.

Students predict the total volumes in each case, and then observe what actually happens when these procedures are performed.

The final stage asks them to explain what they have observed—particularly if what they have observed is different to what was predicted. In this case, the first two test tubes result in 100 ml of the substance. In the third instance, the resulting volume is typically between 97 and 98 ml. What has happened to the other 2 ml and how could it be explained?

In order to pursue a change in thinking, Posner and colleagues (1982) propose three important elements necessary for supporting a shift to the new concept. The new concept needs to be: intelligible, plausible and fruitful. For a new concept to be intelligible, it must in some way make sense and be able to explain an experience so that new possibilities might be grasped. If the new concept is intelligible, then the next test involves the need for it to be plausible—it must have the ability (or possibility) to resolve the initial dissatisfaction as well as carry a sense of consistency with other 'knowledge' about the concept. Finally, conceptual change is possible if, when put to the test, the new concept is fruitful—that is, it works, and can actually resolve the issues associated with the original conception. Fruitfulness can become even more helpful if the new conception can go beyond the original and extend and open up inquiry into new ways of functioning/explaining situations.

In terms of learning, it is important to recognise that the conceptual change model describes the accommodation, not assimilation, of a concept. Posner and colleagues (1982) argue that assimilation hinges on the use of existing concepts to explain new phenomena, whereas accommodation is about situations in which new phenomena cannot be assimilated, so replacement is initiated. Thus conceptual change occurs through accommodating the new phenomena and pushing out (or replacing) the previous (alternative) conception. See Chapter 2 for further explanations and illustrative examples of this construct.

In the above example, if students think that matter is made of 'stuff' where you just have more or less water or methylated spirits in what is known as the **continuous model**, then they will have difficulty in explaining the 'missing 2 ml'. If they have a notion of particles that can be packed in different ways, then they are likely to have more chance of explaining the 'missing 2 ml', as the different particles pack differently in the mixture of water and methylated spirits.

One of the implications for teaching based on a conceptual change model is the importance of creating a sense of dissatisfaction or cognitive conflict in relation to a student's existing alternative conception. In terms of teaching, this assumes that a teacher possesses ways of

continuous model the alternative conception students often hold of matter as smooth and infinitely subdivisible

'bringing out' their students' alternative conceptions and has ways of working with their prior knowledge in order to create a sense of dissonance in their thinking. And herein lies the challenge for teachers. If such cognitive conflict/dissonance can be created, then the pedagogical experiences that follow need to build appropriately on the ideas of intelligible, plausible and fruitful so the new conception is accommodated.

In thinking about science learning in relation to alternative conceptions, it is clear that there are major implications for understanding the nature of science teaching. For example, if a teacher of science is unaware of students' alternative conceptions, it stands to reason that it is unlikely that their practice will be designed purposefully to address those alternative conceptions. If that is the case, a possible outcome is the well-recognised 'school science' learning outcome—when students apply their 'school science' (usually as propositional knowledge) to answer questions in tests, but revert to their alternative conceptions for explaining the real world. In essence, 'school science' learning may have little real-world application, and so does not help students to explain everyday phenomena in accord with the 'correct' science explanation of the time. Clearly, that is not the type of learning outcome that teachers of science would want for their students.

It is not hard to see how complicated science learning and teaching become when a teacher begins to push back against transmissive teaching based on the 'telling as teaching' and 'learning as listening' approach to practice. Importantly, just as cognitive dissonance may be a catalyst for change in students' thinking, the same applies when it comes to influencing teachers' conceptions of practice and the development of expertise. The notion of PCK offers a way of conceptualising what it means to develop expertise in practice because it acknowledges and values the teaching–learning relationship with specific regard to the subject-matter.

PEDAGOGICAL CONTENT KNOWLEDGE

Pedagogical content knowledge (PCK) describes the amalgam of subject-matter knowledge and pedagogical knowledge that teachers develop in their specialist area of teaching—for example, as a specialist teacher of science.

What is PCK?

PCK offers insights into the specialist's knowledge, skills and ability because it illustrates how what is taught is based on understandings of teaching and learning that go beyond an 'activities that work' approach to classroom practice. It is about knowing how to develop and differentiate between creating classrooms that keep students busy (and may be seen as fun) and classrooms in which students are genuinely engaging in their learning of science. PCK is about teaching specific content in a particular way, for a particular reason, with a particular group of students. PCK offers insights into how specialist teachers think about and shape their practice in response to their specific pedagogical context. With an increasing emphasis on primary teachers also having some specialist teaching expertise, PCK becomes an interesting idea for all teachers to consider.

Lee Shulman (1986) introduced the notion of PCK, and it immediately resonated with educators because it offered a way of understanding what it is that specialist teachers know and are able to do. In the case of science teaching, PCK illustrates why a teacher's knowledge of, and approach to, teaching science is sophisticated business, and sheds light on what it means to be an expert teacher (Loughran, 2010).

Shulman put forward PCK as an abstract concept that carried an understanding of teaching that comprised so much more than simplistic views of teaching as training, or thinking that teachers' expertise was dependent on content knowledge alone—in other words, that content knowledge is all that is required to be a 'good' teacher. Shulman argued that a specialist, such as a science teacher, understood subject-matter content in a way that was shaped by how students learned that content. Therefore, science teachers' PCK was evident in the way they understood science concepts through the lens of teaching in a way that meant their practice transformed that content through learning, as opposed to simply delivering information to be committed to memory.

It is not hard to see that the notion of PCK challenged transmissive views of teaching—'telling as teaching' and 'listening as learning'. PCK helps to make clear that expertise goes beyond the 'what' of subject-matter and knowledge and centres on the 'how' and 'why' that dramatically impact learning. PCK offers a clear challenge to transmissive views of teaching, or what Paolo Freire (1972) describes as the **banking model**. PCK is important because it draws attention to science learning as a major shaping force in thinking about the ways in which a teacher develops, structures and enacts their teaching of science for understanding.

In the earlier vignette, the ways in which pedagogy and subject-matter content interact to shape the pedagogical experience illustrate how PCK may begin to emerge and be recognised as a key aspect of a teacher's specialist knowledge of teaching.

banking model Freire's description of teaching dominated by an approach to rote learning and the absorption of information

PCK is an interesting idea because, as Case Study 8.1 suggests, there is a major difference between using a role-play as an 'activity that works' and using it because the role-play helps to make the particular content make sense for students; making the abstract concrete in this case is recognition of an issue with learning about the specific content. Because the role-play purposely complements issues about learning the subject-matter, it offers an insight into PCK, with content and the pedagogy being used together to shape learning in new and powerful ways (the 'why' of practice).

Case Study 8.1 is a brief example of one classroom situation. Exploring how the development of PCK might be possible at a level beyond a single situation offers opportunities to uncover and illustrate how PCK can shape the development of teachers' knowledge of teaching at a higher level.

Making PCK useable

The research literature highlights a number of ways in which PCK has been taken up and studied in science education (Gess-Newsome, 2015). In many cases, the range of PCK studies illustrates how PCK has been researched across different areas of science, but generally the research literature speaks to a research audience much more than it does to a teacher audience. However, putting the concept to use in practice is an important measure of value for teachers, and therefore matters when translating research ideas into classroom knowledge and practice.

For PCK to be meaningful and applicable in the teaching of science, real-world examples of how it might be captured and portrayed are required. One way of doing that is described through the use of content representation (CoRe) and the associated Professional and Pedagogical-experience Repertoires (PaP-eRs), examples of teaching in action (see Loughran, Berry and Mulhall, 2012 for a full description).

> CoRes offer [a] most useful technique . . . for eliciting and recording PCK directly from teachers. The method is clearly centred in teachers' skills and knowledge, so a completed CoRe provides a powerful means of recording the work of an experienced teacher available for sharing and exemplifying good practice. However, the method is not unproblematic: the questions posed in a CoRe are challenging . . . the PaP-eRs strategy offers an

> [additional] method of [further] probing PCK [through windows into practice]. (Kind, 2009, p. 195)

In developing CoRes, teachers are asked to think about a big idea in science and then respond to a series of questions designed to probe teachers' pedagogical purposes and reasoning regarding the big idea. The CoRes represent a summary of teacher thinking and expertise about this big idea, while the PaP-eRs attempt to capture the nuances that exist between different teachers.

A CoRe is designed as a matrix (see Table 8.1 for an example of an abbreviated CoRe) with the important science ideas/concepts (big ideas of the unit) across the top and a number of pedagogical prompts that probe each of the big ideas down the side. There is no defined number of science ideas/concepts for a topic, but research suggests that approximately five to eight commonly capture the essence of thinking about a topic.

As Table 8.1 illustrates, the nature of a big idea is very different from listing a 'piece of scientific information'. For example, 'matter is made up of small bits called particles' is a very different way of thinking about matter in contrast to simply stating that 'particulate matter comprises atoms and molecules'. Developing big ideas for a topic is initially difficult, as thinking conceptually about a topic is different from listing the pieces of information that make up the 'knowledge' of the topic. In the example in Table 8.1, the big idea of 'a force is a push or a pull' is actually a concept developed in the primary years of education. As the CoRe indicates in Table 8.1, this is for teaching at Year 10 level, where the added sophistication comes from the secondary idea of, for example, 'a force on an object is external to the object'.

The eight prompts (left-hand side of the CoRe) explore the pedagogical reasoning that underpins the ways in which the big ideas are understood and explored in practice. The first prompt, 'What you

Table 8.1 An abbreviated CoRe illustrating the big ideas (horizontal axis) and pedagogical prompts (vertical axis)

	Important science ideas/concepts			
This core is designed for Year 10.	**A:** **A force is a push or pull.**	**B:** **Objects distort when forces are put on them. That causes them to exert a force on the object that distorted them.**	**C:** **Objects that are stationary always have an equal balance of forces acting on them.**	**D:** **Frictional forces are caused by distortion of two surfaces sliding over each other.**
1. What you intend the students to learn about this idea.	A force is a push or a pull only. A force on an object is external to the object.	Reaction forces (usually) occur because an object distorts (as a result of a force acting on it).	If an object is at rest, then it must have balanced forces acting on it—that is, the net force is zero.	Frictional interactions result in real forces and are due to tiny bumps on surfaces.
2. Why it is important for students to know this.	The teacher must establish this first or risk students making alternative constructions of everything else.	The distortion explanation is important; otherwise reaction forces merely seem to be an artificial convention.	This is important for understanding one aspect of Newton's first law (i.e. an object at rest will remain at rest unless acted upon by a net force).	

	Important science ideas/concepts			
This core is designed for Year 10.	**A:** **A force is a push or pull.**	**B:** **Objects distort when forces are put on them. That causes them to exert a force on the object that distorted them.**	**C:** **Objects that are stationary always have an equal balance of forces acting on them.**	**D:** **Frictional forces are caused by distortion of two surfaces sliding over each other.**
3. What else you know about this idea (that you do not intend students to know yet).	In scientific terms, a force produces accelerated motion; however, that would be a poor way to start teaching about force.	The idea of distortion is enough: there isn't a need to draw out the notion of action–reaction pairs.	Early on, avoid field forces. A key idea when fleshing out the forces on an object is that when object A exerts an (action) force, there is a (reaction) force back in the opposite direction on object A.	
4. Difficulties/ limitations connected with teaching this idea.	The idea that a force is a push/pull is NOT tricky BUT the idea that the force on an object is external to an object IS often problematic for students.	It is hard to believe that rigid objects distort.		

This core is designed for Year 10.	Important science ideas/concepts			
	A: **A force is a push or pull.**	**B:** **Objects distort when forces are put on them. That causes them to exert a force on the object that distorted them.**	**C:** **Objects that are stationary always have an equal balance of forces acting on them.**	**D:** **Frictional forces are caused by distortion of two surfaces sliding over each other.**
5. Knowledge about students' thinking that influences your teaching of this idea.	Sometimes it is the terminology that causes a problem rather than the concept itself. Finding ways of drawing the distinction matters.	Most students don't hold a notion of distortion despite having lots of experience of distortion—for example, trampolines, springs		
6. Other factors that influence your teaching of this idea.	Interpretive discussion facilitated by the teacher is particularly valuable in this area as students' prior knowledge and experiences mean they are often better placed to challenge and convince one another.	It can be too easy to simply teach the ideas in the unit as facts; however, this does not help develop students' understanding.		

This core is designed for Year 10.	Important science ideas/concepts			
	A: **A force is a push or pull.**	**B:** **Objects distort when forces are put on them. That causes them to exert a force on the object that distorted them.**	**C:** **Objects that are stationary always have an equal balance of forces acting on them.**	**D:** **Frictional forces are caused by distortion of two surfaces sliding over each other.**
7. Teaching procedures (and particular reasons for using these to engage with this idea).	Defining the boundaries: draw out alternative uses of the word 'force'. The example of signs displaying 'No animals on freeway' is useful for distinguishing between scientific meaning of the term animal and the common everyday meaning used.			
8. Specific ways of ascertaining students' understanding of or confusion around this idea.	Listening for appropriate use of words. The absence of inappropriate use of 'force' indicates students have grasped the idea that in science we use 'force' to mean a 'push' or 'pull'.			

Source: Loughran, Berry & Mulhall (2012).

intend the students to learn about this idea', is a beginning point for unpacking understandings of what matters in learning the particular content.

The next prompt, 'Why it is important for students to know this', begins to question how the work is relevant (or might be made relevant) to students' everyday lives. Over time and through experience, teachers often develop more sophisticated ways of linking concepts to 'students' need to know, and this prompt helps to make those ideas and actions explicit.

'What else you know about this idea (that you do not intend students to know yet)' begins to explore aspects of the level of complexity a teacher sees as important to uncover with students about the topic. It also offers insights into a teacher's thinking about how to manage the tension between over-simplification and creating unnecessary confusion, which leads to the next prompt, 'Difficulties/limitations connected with teaching this idea'. This prompt is a reminder of such things as alternative conceptions (as noted earlier in the chapter); the nature of concepts in relation to the issues such as being abstract and needing to be made concrete (as in Case Study 8.1); the value and limitation of analogies, models and metaphors; and so on. This prompt goes to the heart of how a teacher genuinely considers learning about an idea and how teaching needs to be responsive to those issues if understanding the concept(s) is genuinely to be developed.

The next prompt, 'Knowledge about students' thinking that influences your teaching of this idea', builds on the previous prompt, but is designed to dig more deeply into a teacher's experience of how students usually respond to the topic and how their teaching is structured in response to that thinking.

'Other factors that influence your teaching of this idea' is a prompt designed to begin to unpack a teacher's knowledge of the contextual factors about the situation and to shed light on how students respond to those features of the teaching–learning environment and how that shapes the nature of the teaching that occurs in that particular environment.

The next prompt, 'Teaching procedures (and particular reasons for using these to engage with this idea)', is designed specifically to highlight a teacher's pedagogical reasoning and goes to the heart of the difference between an activity that works and the use of a teaching procedure for a particular reason. This is a feature of PCK that highlights why a teacher's knowledge of practice is important beyond the level of 'activities that work'; it links teaching and learning in meaningful ways and makes clear what it means to develop purposeful pedagogy. The final prompt flows naturally from a consideration of teaching procedures.

'Specific ways of ascertaining students' understanding of or confusion around this idea' is the final prompt, and is based on how to apprehend and monitor students' understanding and progress (both formally and informally). This prompt is designed to explore how teachers approach this aspect of their teaching of the topic in order to gather different perspectives of the effectiveness of their teaching and the ways in which student learning about the topic develops and changes. Although the prompt could be read as being about assessment, it is specifically worded in terms of 'ascertaining students' understanding' in order to ensure a broader interpretation than might be the case if viewed only as an approach to summative assessment. The prompt therefore illustrates how PCK includes feedback on teaching and learning in ways that inform practice, rather than just measuring outcomes.

A CoRe is not intended to contain a prescribed amount of information or ideas; nor is there one correct version of a CoRe—clearly the nature of the big ideas shapes the response to the prompts in different ways. When developing a CoRe, some of the cells may contain more detail than others, and some may well be left empty (all depending on the knowledge, experience and expertise of the teacher(s) completing the CoRe). As the research illustrates, different groups of teachers develop different CoRes for the same topic and are influenced by their thinking about the 'way in' to the topic. A CoRe provides insight into a teacher's experiences of practice with the given content/topic.

Supporting the CoRe are the PaP-eRs, which are designed to offer glimpses of science teachers' practice, thinking and understanding of the particular content in particular ways at particular times. There may be many PaP-eRs associated with a CoRe, each of them designed to highlight particular aspects of the CoRe. For example, the following PaP-eR shows how a teacher put 'into action' the prompt 'other factors that influence your teaching of this idea' when working on the topic of force. In focusing on this aspect of the concept of force, it is clear that while the big ideas may appear simple, their use in the everyday is not.

CASE STUDY 8.2: Does gravity need air?

Students' ideas about the relationship between gravity and air are often very different from those accepted by scientists. This PaP-eR explores one teacher's awareness of the misconceptions of his Year 10 students about these concepts and how he endeavoured to make plausible for them the notion that gravity can exist in the absence of air.

As part of their work in a unit about force, I ask students to keep a 'thinking page' where they write questions or ideas that they have about forces that come up from their work in class, or other related experiences. I find that learning about this topic typically raises lots of questions for students as they begin to link their school science learning with their experiences outside class. By regularly collecting their 'thinking pages', I can monitor students' questions and concerns and more closely shape the curriculum around their particular needs. For instance, last week Sam, one of the quieter students in the class, wrote this question on his 'Forces thinking page': Mr Dyson, does gravity need air to work?

This question struck me as one with which students often struggle. Sam's question also prompted me to think about other confusions about gravity and air that are commonly held by students. These include:

- Where there is no gravity, there is no air (e.g. there is no gravity in space).
- Gravity only exists on earth.
- As objects fall toward the Earth (or moon), they get faster and faster because gravity increases.

In helping students to learn about forces, it is important that they understand that gravity is not related to air and, more broadly, how the concept of gravity allows for the explanation of everyday experiences such as falling objects and weight.

After reading Sam's question, I decided that in the next lesson I would use an approach to challenge students' thinking about the relationship between air and gravity. Here's what I did. First, to encourage students to think through their ideas about air resistance, I held a flat sheet of paper at shoulder height and asked them to predict how long they thought it would take for the paper to reach the floor after I let it go. (I have found that asking students to predict the outcome of a demonstration helps to increase their engagement with the ideas considerably. In this case, their prediction also sets the scene for what happens next.)

After a brief discussion of their predictions, I dropped the paper and compared their predictions with how long it took for the paper to hit the ground. I then asked the students to offer their ideas about how I might increase the rate of fall, using the same piece of paper and dropping it from the same height. They quickly predicted that crumpling the paper into a tight ball would achieve this result. We tried this a few times, as different students wanted to test their paper-crumpling abilities. In this way, I set the scene for talking with the class about how changing the shape of the paper changed the paper's rate of fall. Through our discussion, I wanted students to recognise the following things:

- Air affects the way things fall.
- Air gets in the way as things fall.
- Air provides some resistance to things falling because of collisions between the air particles and the surface of the thing that is falling.
- Increased surface area means increased resistance (more collisions) as things fall through the air.

In my previous approach to teaching these ideas about gravity and air, I would simply have 'told' the students about the relevant science. However, more recently I have come to see that there is more to shifting students' thinking than simply telling them what to think. For meaningful learning to occur, students need to have experiences where they can begin to articulate new ideas for themselves. This will support learning when more complex ideas about forces are developed later in the unit (for instance, 'freefalling' and terminal velocity).

Following our class discussion, when I felt convinced that most students were comfortable with these four ideas, I introduced a second demonstration. This time, I placed a coin and feather together into a glass tube and inverted the tube. Then, using a vacuum pump, I evacuated the air from the tube and repeated the demonstration. Before inverting the tube each time, I asked the students to predict any difference in the rate of fall between the two objects. (This usually presents no problem for students when the air is present and the tube is inverted. However, they are often fascinated to see that, when the air is removed from the tube, the two objects fall at the same rate.) The purpose of this demonstration is to help students to recognise that: (1) things fall when air is not present; (2) the rate of fall is the same for all objects when air is not present; and (3) air causes things to fall at different rates.

Elaborating these ideas further, I showed a short video clip of the Apollo 11 moon mission, in which one of the astronauts releases a heavy object (a 1.32 kg hammer) and a light object (a 0.03 kg feather) simultaneously from approximately the same height and allows them to fall to the moon's surface. (The objects fall together and strike the surface simultaneously.) After viewing the video clip, I asked students to explain what they had seen and to link it to the ideas they had encountered about air and falling. From these class activities, I wanted to feel confident that they had grasped the ideas that objects fall at the same rate on Earth in the absence of air; objects fall at the same rate on the moon; and the rate of fall on the moon is less than the rate of fall on the Earth. Listening to students explain their thinking to me and to each other is an important means for me to monitor their understanding of these concepts. (This also includes listening for the common confusions I mentioned earlier.)

However, even after the demonstrations, video and discussion, there is still an unresolved problem which some students will pick up on. So far, we have really only dealt with the effects of air resistance. Although I have been able to provide students with convincing evidence of the effects of the presence or absence of air on falling objects, I have not yet dealt with what gravity is or how it works. Unlike air resistance, such ideas are not readily understood through practical demonstration or teacher explanation. In fact, the nature of gravity is something that scientists also find difficult. Gravity is not currently well understood by the scientific community, although it is a useful construct that enables scientists to predict some of its effects in different situations.

So I offer the following explanation of gravity to the students. Every object in the universe seems to attract every other object with a force that is related to the size of their masses. We call this force 'gravity' (although scientists aren't quite sure what gravity is). Because gravity is only a very weak force, we only notice it if at least one of the objects has a huge mass (such as the Earth). I tell students that I cannot give them tangible evidence of what gravity is in the same way that I was able to demonstrate the effects of air resistance.

In the past, this has led to an interesting diversity of student responses about what kind of knowledge science can offer and the different ways in which science knowledge is constructed and developed (e.g. some students are convinced that science offers conclusive answers to all questions; others are satisfied that science does not offer 'answers to everything'). I emphasise to students that although there are some things that scientists don't understand well, these ideas are still useful to explain why certain things happen. Gravity is an example of such a phenomenon.

I think is it important for students to begin to grapple not only with knowledge of science concepts, but also with ideas about the nature of scientific knowledge. Knowing about how science (and scientists) work is something we are then able to revisit in different units of work.

Following the day's demonstrations and activities, I again picked up Sam's 'thinking page', and this time found the following note: 'Today, Mr Dyson showed us that air affects how things fall, but he wasn't very helpful on how gravity actually works.' (Loughran, Berry & Mulhall, 2012, p. 153)

Together, then, a CoRe and its associated PaP-eRs offer a concrete example of what PCK might look like and how that teaching might be enacted in practice. Because it is particular to the learning and teaching of a particular unit or topic of science, it illustrates the amalgam of content and pedagogy that is at the heart of the notion of pedagogical content knowledge.

Kind (2009, in the direct quote that began this section), like many others who have worked with CoRes and PaP-eRs, recognised that developing a CoRe is not easy because it requires two major shifts in thinking about conceptualising subject-matter and teaching. The first shift is from subject-matter as propositional knowledge (content to be delivered) to conceptual understandings that comprise the big ideas of the content. The second shift relates to recognising the value and purpose of questioning the subject-matter knowledge from an informed and strategic pedagogical perspective.

PaP-eRs tend to attract less attention from teachers as they typically focus much more on the development and use of a CoRe as a helpful approach to thinking about a unit of work. Research illustrates that teachers find CoRes helpful for curriculum development, as they differ markedly from more traditional approaches to curriculum documentation.

PCK: Creating a vision for future professional learning

There continues to be considerable debate around what it means to develop as an expert pedagogue. Much of the debate can be traced back to public views of how teaching is understood and interpreted—that is, transmissive approaches to teaching as opposed to teaching of understanding. Transmissive views of teaching tend to see the 'development' of practice in ways that accord with 'doing things to teachers', and are often described as forms of professional development. More often than not, teaching for understanding relates to ways of supporting teachers in taking responsibility for, and control of, their learning about practice in ways that make a difference for their teaching and their students' learning. Therefore, professional learning is qualitatively different from professional development.

The language of professional development is often based on the idea that there is a solution (to a given problem) available, and that by 'up-skilling teachers' or mandating the use of the solution in their practice, the problem will be solved. On the other hand, professional learning is based on a different understanding of development. Professional learning, as it is described by Groundwater-Smith and Mockler (2009, p. 13), can be described as:

- focused upon transformation, of both self and school
- collaborative, with opportunities for teachers to build authentic collegiality
- ongoing rather than solely project-based
- capable of engaging teachers in creating knowledge about and for practice
- encompassing of opportunities for teachers to develop and hone their professional judgment.

CONCLUSION

The development of expertise revolves around the nature of the professional judgements that teachers make in organising and structuring their teaching in response to their understandings of their students' learning needs, issues and concerns. Those professional judgements are continually informed by the nature of a teacher's experiences, knowledge and practice, all of which combine to shape the quality of the pedagogical experiences they create for their students.

PCK is one key aspect of teachers' professional knowledge. It highlights the knowledge, skills and abilities of the specialist (science) teacher. PCK is not about a set of rules to be learnt and applied; rather, it is about the professional judgements that inform a teacher's practice.

PCK is developed as a consequence of professional learning over time. Creating professional learning opportunities is one way in which education systems can help to focus serious attention on PCK, which in turn can foster the development of expertise. PCK offers a vision for what it means to be able to articulate knowledge of teaching and learning, and how to apply that learning in practice in ways that make a difference. In so doing, it also offers an overt example of pedagogical expertise.

DISCUSSION QUESTIONS

8.1 Can you (and some colleagues) think about a big idea and develop your own CoRe? (As pre-service teachers, you may want to focus on prompts 1, 2, 3, 4 and 8.)

8.2 In developing the CoRe, what sort of discussion took place? Would you be able to capture this as a PaP-eR?

8.3 How did you identify a big idea?

8.4 What does the construct of PCK mean to you at this stage of your teaching career? How might you draw on PCK to inform and enhance your practice?

8.5 With your responses to these questions in mind, why is conceptual change such an important consideration in terms of how we approach science learning and teaching?

FURTHER READING

Loughran, J., Berry, A. & Mulhall, P. (2012). *Understanding and Developing Science Teachers' Pedagogical Content Knowledge*, 2nd ed. Rotterdam: Sense.

REFERENCES

Freire, P. (1972). *Pedagogy of the Oppressed*. New York: Herder & Herder.

Gess-Newsome, J. (2015). 'A model of teacher professional knowledge and skill including PCK'. In A. Berry, P. Friedrichsen & J. Loughran (eds), *Re-examining Pedagogical Content Knowledge in Science Education*. London: Routlege, pp. 28–42.

Groundwater-Smith, S. & Mockler, N. (2009). *Teacher Professional Learning in an Age of Compliance: Mind the Gap*. Dordrecht: Springer.

Kind, V. (2009). 'Pedagogical content knowledge in science education: Perspectives and potential for progress'. *Studies in Science Education*, 45(2), 169–204.

Loughran, J.J. (2006). *Developing a Pedagogy of Teacher Education: Understanding Teaching and Learning About Teaching*. London: Routledge.

—— (2010). *What Expert Teachers Do: Teachers' Professional Knowledge of Classroom Practice*. Sydney: Allen & Unwin.

Loughran, J., Berry, A. & Mulhall, P. (2012). *Understanding and Developing Science Teachers' Pedagogical Content Knowledge*, 2nd ed. Rotterdam: Sense.

Posner, G.J., Strike, K.A., Hewson, P.W. & Gertzhog, W.A. (1982). 'Accommodation of a scientific conception: Toward a theory of conceptual change'. *Science Education*, 66(2), 211–27.

Shulman, L.S. (1986). 'Those who understand: Knowledge growth in teaching'. *Educational Researcher*, 15(2), 4–14.

Wandersee, J.H., Mintzes, J.J. & Novak, J.D. (1994). 'Research on alternative conceptions in science'. In D. Gabel (ed.), *Handbook of Research on Science Teaching and Learning*. New York: Macmillan, pp. 177–210.

Part IV

New approaches in teaching and teacher development

As stated in the preface, an intention of this book has been to extend your thinking about what matters in science education by planting seeds of ideas for you to revisit and reinvigorate in the future as you establish yourself as a teacher of science. This includes thinking about the development of your professional practice and identity over your career, as well as how you might leverage other science-rich contexts (such as informal learning settings) to further enrich your teaching of science and your students' learning of science. With the growing prominence of STEM education, there is an opportunity to consider how you fit into this space as a teacher of science and what influence it might have on the practices you adopt.

Part 4 teases out three key areas for consideration:

- *professional learning and growth*—thinking about your transition into professional practice and the opportunities on which you can draw over your career to develop your expertise and enhance learning
- *informal learning settings*—raising awareness of the role that informal settings can play in science education through the promotion of authentic and innovative approaches to science
- *STEM education*—highlighting the importance of STEM education for the future workforce and drawing out the inherent twenty-first century learning skills to inform practice.

This section of the book is valuable, as it provides practical guidance and insights into the ways in which you might choose to extend your thinking about and enhance your practices in science education into the future.

CHAPTER 9

TRANSITIONING INTO PROFESSIONAL PRACTICE

KATHY SMITH *and* **REBECCA COOPER**

LEARNING OUTCOMES

After reading this chapter, you should be able to:

- appreciate that teaching is characterised by ongoing and shifting dilemmas
- appreciate that developing expertise in science teaching takes time and is challenging work
- understand that opportunities for personal learning and professional growth will exist in your teaching practice, and throughout your career
- understand that teachers demonstrate expertise when they are able to use their professional knowledge to effectively enhance student learning.

> [Science teaching is] like when you learn how to drive for the first time and you need to put all your effort into the clutch, gears and accelerator. However, as your learning progresses and you become more experienced, those separate tasks combine, become automatic and then you start to pay attention to other things. (Speakman, 2012, p. 20)

While comparing teaching and driving may seem a strange—even incompatible—analogy, by reflecting on these two life events, this science teacher (Speakman) has identified 'experience' as a key enabler of knowledge development. Experience assists a driver to become familiar with road events and repeatedly process lots of sensory input to develop safe and purposeful driving behaviour. While driving is initially extremely difficult, in time it becomes less about coordinating various physical reactions and more about a capacity to read conditions and anticipate and respond effectively to a range of changing traffic events. In a similar way, initial teaching experience enables familiarity with many everyday classroom routines. As a consequence, teachers develop the knowledge and skills necessary to effectively attend to everyday tasks such as managing timetables, learning students' names, planning a lesson and dealing with curriculum. In time, knowing how to effectively manage what become familiar aspects of teaching allows a teacher to focus attention on other more complex areas of their work—for example, finding ways to effectively monitor, support and enhance individual student learning while also attending to the needs and interests of groups of students. A number of similarities are at play in both scenarios: experience affords opportunity to understand predictable events in ways that require less concentrated effort; familiarisation involves noticing, drawing upon and weaving together the similarities and resemblances of experience; and, potentially, time enables the development of an improved capacity to anticipate, accurately read and then appropriately respond to a complex mix of even more challenging events.

The beginning teacher may easily under-estimate the amount of time and persistence involved in developing the professional knowledge that underpins quality teaching. Expert teachers make teaching look simple, as they appear to seamlessly anticipate likely scenarios and provide support and opportunity for rich and meaningful student learning. Yet what is at play in expert teaching is careful and considered reasoning, because expert teachers make sense of and respond purposefully to a range of teaching events. Developing such professional knowledge about teaching takes time, and is the result of a willingness to:

- acknowledge a need to seek understanding
- notice and interrogate experience
- persist through challenges
- critically reflect on actions and thinking.

Professional competence and growth occur when teachers are willing to explore their own assumptions (Brookfield, 1995) and take time to understand more about why they work in the ways they do. Expert teachers draw from their experience to develop new thinking and understandings about learning and teaching, and demonstrate this knowledge in contextually relevant ways (Smith, 2017a).

This chapter will encourage you, as a beginning teacher, to appreciate science teaching as challenging work characterised by ongoing and shifting dilemmas. The information discussed will encourage you to consider how this context shapes your professional growth, in particular how such growth ultimately depends on your willingness to see the importance of constantly paying attention to your teaching practice and also the beliefs, values and personal assumptions you hold that shape your thinking about learning and teaching. Professional growth takes time; therefore, the chapter encourages you to be patient while at the same time encouraging you to recognise that your teaching context provides a rich opportunity for relevant and useful professional learning. As you work to learn more from your teaching experiences, you will inevitably face many challenges; however, while opening up opportunities for learning, these will undoubtedly produce feelings of uncertainty as well. This chapter highlights that these feelings are a valuable part of the learning process and when strategically examined, through critical reflection, the insights gained will potentially build your capacity to make increasingly sophisticated and well-informed pedagogical decisions about the conditions that support quality learning.

BEGINNING A JOURNEY TO EXPERTISE

Remembering how the complex became so familiar is the way Speakman (2012) remembers her early experiences of science teaching. Teaching science required her to master what initially appeared to be an overwhelming and disparate range of considerations, responsibilities and responses. Science teachers are required to pay attention to many facets of professional practice, such as:

- the complexity of student–teacher relationships
- students' cognitive capacity for conceptual development
- the nature and importance of conceptual knowledge in a range of science areas
- the demands of government accountability in terms of curriculum expectations
- the expectations of society as a whole.

These are merely a few of the many considerations involved in teaching science. For this teacher, a combination of experience, assistance from teacher mentors and her capacity for critical thinking made it possible to develop the specific knowledge and skills she needed to teach science. Such learning also built her capacity to identify what she valued in relation to learning and teaching. In time, she became better equipped to align this thinking with her teaching action, enabling her to weave such competencies within and around all that defines school operations, being a teacher of science and promoting student learning. Her knowledge of science teaching is now so deep that she no longer recognises these elements as separate tasks, but rather as interrelated ways of knowing, resulting in actions that have become 'automatic'. This familiarity now allows her to 'pay attention to other things', such as creating conditions that enhance learning for every student.

Speakman's comments serve as a powerful reminder for pre-service teachers that it takes time and experience to move beyond simply 'doing' teaching to arrive at a career point defined by a deep 'understanding' of teaching. While the practice of teaching will never become something that is governed by rules as predictable as those related to driving, experience does play an important role in quieting the noise and practical demands of the classroom. Experience provides opportunity for a beginning teacher to do more than manage tasks, so that they become more effective at noticing and effectively responding to the diverse learning needs and interests of students. This is the beginning of the journey to expertise, teachers of science value and actively attend to the challenges present in their teaching to explore and develop the thinking and skills they need to effectively deconstruct complex conceptual ideas and find ways for students to build meaningful knowledge. This willingness to interrogate practice, and to think and work differently, is the hallmark of expert teachers of science. These are the teachers whose practice appears to seamlessly unpack and then weave together all the complexities of learning in ways that seem almost automatic. These teachers make teaching science look easy, although the reality is far different.

Mastering the mechanics and awareness required to achieve smooth and comfortable driving shares some similarities with developing purposeful teaching. Both are about personal awareness and working to build a capacity to manage change. Yet learning about teaching moves beyond the parameters that define learning to drive because teachers must find ways to successfully attend to the ever-changing educational needs of the individual. They do so within a complex system of learning and deciding what matters while holding all of these considerations in alignment and balance. This is really difficult work, and is intellectually and emotionally demanding. Quality teaching is determined by how a teacher thinks about learning and teaching, and the relationship that exists between these intents: **pedagogy**. What teachers think and what they believe ultimately shapes the kind of learning young people get (Hargreaves, 1994). So what role does teacher thinking play in developing expertise in teaching?

pedagogy the relationship between learning and teaching

QUALITY TEACHING AND TEACHER THINKING

Walking into any classroom on any given day, it would be easy to assume that science teaching is largely about selecting activities and managing student behaviour because that is what can be seen. Yet there is a great deal of other activity taking place, most importantly inside the teacher's head as they make decisions about what to teach and why. Teachers require a high level of professional expertise to produce quality learning and teaching, and experienced teachers of science operate from a complex knowledge base (Corrigan & Gunstone, 2011) yet rarely articulate the thinking that drives their decision-making, and they often find it difficult to explain the knowledge that has become so familiar and practised in their work. This knowledge of practice, first identified in research as teachers' practical knowledge (Clandinin, 1985) and more recently referred to as teacher knowledge of practice (Loughran, 2010), guides the pedagogy of effective teachers. Some examples of aspects of this professional knowledge include teachers knowing: curriculum content; their students and how they learn; how to make sense of and attend effectively to the factors directly influencing student thinking (such as alternative conceptions—see Chapter 2); diverse linguistic, cultural and religious backgrounds; and how to find ways to draw on student experience to enhance learning. Very importantly, a teacher's personal aims, purposes and the values that underpin and frame their particular approach to science education are also at play here, and form part of this professional knowledge. A teacher's knowledge of self-identity and their sense of personal adequacy are also essential aspects that define the 'personal' in professional practice and greatly influence the type of learning experienced by students. Expert teachers of science draw from this complex knowledge of practice in ways that effectively address the physical, social and intellectual development of all students while also attending to the individual differences of each learner.

Teaching science involves far more than what a teacher can be seen to do: it is not merely a technical activity, but complex. While expert teaching is always purposeful, it can also be uncertain and problematic.

Teaching isn't rocket science—it's harder!

Developing the level of expertise required to become practised and proficient in nurturing meaningful science learning is not only challenging; it is at times all-consuming. Experience provides opportunity for teachers of science to determine what makes a difference to their students' learning and, through experience and reflection, teachers themselves become active lifelong learners, continually examining their practice, identifying the strengths and weaknesses of the approaches they adopt and recognising the assumptions that frame their work in the classroom (Brookfield, 1995). As Fuller (2013) suggests, successful teachers are individuals who are able to orchestrate and conduct a 'symphony of human development'.

Yet teaching is not an isolated activity: decisions about learning and teaching are situated within a prevailing educational culture that constantly demands perfection, compliance and accountability. The systems in place to support formal school-based learning are complex (Opfer & Pedder, 2011). Teachers of science, like any teachers working within these structures, must develop a capacity to balance personal values and beliefs, professional growth, the needs of the individual student, and the complex and fluid nature of learning—all within the often contradictory expectations of standardised and structured learning systems. It is inevitable that challenges and problems will arise for teachers as they teach, as identified by Loughran (2010). Such challenges and problems are not always due to the inadequacy or incompetence of teachers; they reflect the nature of teaching as 'problematic' (2010, p.3). In reality, learning does not happen in a linear or sequential way; rather, it is fluid and often nuanced, and this creates issues and challenges that often confront, confound and disrupt education systems, teachers and students. Therefore, teaching by its nature is problematic because students' learning needs and responses will always be diverse as they explore new ideas, and apply and integrate new knowledge and skills into existing schemas of understanding. Expert teachers of science acknowledge the challenges of this work and accept the need to find ways to effectively read student thinking and respond accordingly; they build relationships, provide support, regularly assess for understanding and provide targeted assistance where needed.

Teaching is therefore not technical, and it cannot be easily explained as a set of predictable routines of procedures and activities; instead, it draws from a complex set of personal and social processes and practices concerning the whole person. Ryan Fuller

(2013), a former aerospace engineer and now a high school teacher in Colorado Springs, recently stated, 'When I solved engineering problems, I had to use my brain. When I solve teaching problems, I use my entire being—everything I have.'

PROFESSIONAL LEARNING AND PROFESSIONAL GROWTH

Aspirations for quality teaching are informed and shaped by a rigorous body of knowledge derived from continuous research. This body of work has shaped and defined teaching as a discipline—a specialised area of work. Teaching is a profession that serves a greater social purpose: to foster learning and the development of the individual. Teaching science develops the individual in terms of specific skills, capabilities and conceptual knowledge, all of which promote and develop critical thinking and a particular way of knowing and acting in the world. Scientific literacy encourages agentive engagement where individuals utilise ideas, ways of knowing and experiences to become discerning and active participants in local and global affairs. Quality science teaching is therefore informed and underpinned by ongoing educational research in many areas, such as:

- student learning
- pedagogical content knowledge
- interactions and learning conditions that promote meaningful learning
- understanding more about the relationship between the teacher's personal knowledge and understandings about the nature of science
- teachers' conceptions of their role and the implications for learning
- the use of discourse in the classroom
- teachers' understanding of and capacity to develop learning goals

- teachers' relationship with the nature and quality of student learning experiences, particularly in terms of the degree of authenticity and connectedness to everyday life. (Bartholomew, Osborne & Ratcliffe, 2004).

To become an effective teacher of science requires specific professional preparation, not only in terms of learning about the accumulated knowledge and skills of practice that characterise this work, but also developing an awareness of the necessity for strict adherence to codes of professional conduct. Respecting, acknowledging and attending to the values and beliefs held by teachers is fundamental to broadening a teacher's identity as both a learner and a professional (Smith, 2017a). As professional educators, teachers of science (inclusive of both primary and secondary teachers) are required to cooperate and collaborate with leadership, peers and students; they pass on an accumulated culture of practice and wisdom and, because of their specialised knowledge, are accorded a degree of professional autonomy in decision-making about important aspects of their work. Yet such professionalism relies on the teacher continuously seeking to develop a level of professional expertise, and this requires a willingness to interrogate personal practice to achieve continuous growth and development.

So how and when do teachers develop such expertise, and what do they need to pay attention to in order to improve their teaching skills and enrich their personal capacity to work within the many complex contextual realities that define teaching? As previously discussed, teacher proficiency and expertise develop over time, through experience and critical reflection (Brookfield, 1995). This expertise has led to a recognition of the critical role teachers play in determining quality teaching and learning (Timperley, 2011). But what does the development of this expertise look like for a practising teacher? Research has been done that explores the possible stages/phases of career for teachers (Day et al., 2007; Huberman, 1989), and this information provides a useful guide for beginning teachers as they attempt to consider their career trajectory.

STAGES OF CAREER—BUT STILL A PERSONAL JOURNEY

> [We need to stress the importance of] working gingerly with a linear, deterministic, highly psychological stage-development model. Not only because of its partial validity, but also because it encourages us to lump individuals together whose trajectories, when examined carefully, have many more discrepancies than commonalities . . . teachers, like other adults, are not simple passive marionettes whose strings are being pulled . . . (Huberman, 1989, p. 358)

Huberman (1989) identifies seven stages in the career of a teacher that are useful when considering the possible shifts that teachers make throughout their careers (see Table 9.1). In a similar way to there being stages of learning to drive (L plates, P plates and full licence), so too there are stages of a teaching career. However, unlike the stages of learning to drive, the stages of career should not be considered as linear, as Huberman (1993, p. 195) found that 'professional career journeys are not adequately linear, predictable or identical—are often in fact unexplainable using the tools at our disposal'. This highlights the personal nature of a teacher's career journey and emphasises the need for due consideration to be given to each teacher's beliefs, values and experiences.

Day and colleagues (2007) reference Huberman's (1993) work and describe six professional life phases of teachers: 0–3, 4–7, 8–15, 16–23, 24–30 and 31+ years of teaching. Similarly to Huberman (1989), Day and colleagues recognise professional life phases as years of experience, as their research suggests that while

Table 9.1 Stages of career and their characteristics

Stage	Characteristics
Career entry First few years	Survival Reality shock Discovery Trial and error Preoccupation with self and sense of inadequacy Pride of having 'place' within a profession Enthusiasm
Stabilisation 2–5 years	Deciding to make a career of teaching Sense of instructional mastery Less concern with self and more with focus on instructional goals Greater confidence in abilities More flexible in the classroom
Diversification and change 5–15 years	Attempts to increase one's effectiveness Awareness of constraints limiting impact on one's students Awareness of inadequate practice, resources and system leads to activism Willingness to take on more responsibility
Stock-taking 12–20 years	Feelings of routine and a need to break out Increased reflection Reviewing one's life and career Consideration of career shift
Serenity 15–30 years	Self-accepting phase Working mechanically but feeling effective Career ambition and level of investment decrease Greater distance between teacher and pupil
Conservatism 30–40 years	Complain about pupils Views change as rarely making a difference Sees young colleagues as arrogant and administration as 'spineless'
Disengagement 35–45 years	Gradual detachment Shift to retirement Generally positive attitude but energies focused elsewhere

Source: Huberman (1989).

the teachers' experiences are not identical, there are key aspects of the professional life phases that are shared by teachers due to their experience and not because of their age. For example, a teacher who commences their teaching career later in life will still experience the same tensions and concerns as a young teacher who has come straight through secondary school and university into the classroom. Your teaching know-how grows and is influenced by your experiences and your time in the classroom. Teachers can improve simply by having a go and by doing something many times over; however, they really learn and develop when they are able to reflect on what they have done and consider how they could improve it for the next time. Are you able to identify teachers with whom you have worked who may be at the different stages outlined in the above table? If so, are you able to identify the strategy they used to continually challenge and improve their teaching practice?

So what does shifting between these stages/phases mean for your teaching practice? How do these shifts occur and what instigates them? Tripp's (1993) work would suggest that many teachers shift from having lessons that are extremely purposeful in the way they are put together to being somewhat automatic:

> Most teachers realise that, although their routines originally have been consciously planned and practised, they will have become habitual, and so unconscious, as expertise is gained over time. Indeed, our routines often become such well-established habits that we often cannot say why we did one thing rather than another, but tend to put it down to some kind of mystery such as 'professional intuition' or simply 'knowing'. (1993, p. 17)

It is possible that you have had an experienced mentor say to you on professional experience, 'Lesson plans are a waste of time,

I don't ever use them, I just know what I'm doing.' Perhaps the same could be said for experienced drivers who no longer feel they need to consciously think about driving because it is something they can just automatically do. Is this mindset okay? Are experienced teachers correct when they say that lesson plans are not necessary? Perhaps such comments may reflect more about reaching a certain career stage characterised by a confidence with content, knowledge and strategies about what to do in the classroom. Yet all teachers need to take time to consider explicitly the effectiveness of the learning conditions they are creating for their students, to ensure that this level of personal comfort equates with actions and approaches that are still appropriately attending to students' learning needs. Professional learning should not slow with experience; rather, experience should provide an opportunity to recognise new ideas and new challenges. This raises some key questions for all teachers:

- While education systems demand accountability, how can a teacher determine the aspects of expertise they value and wish to pursue?
- What can teachers do to become active decision-makers about their own professional learning needs?
- How will you monitor your teaching effectiveness and identify your learning needs?

Shifting between stages and phases of career is a personal journey and, as recognised by Huberman (1993), stages of career are not an exact fit: individuals often run across a couple of stages, and may even get stuck at times. So how will you take ownership of your own professional development so that you keep building your ideas and checking your assumptions about what matters? This may not be easy.

Simos (2011) refers to a national survey indicating that 'almost half of Australian parents who teach their children to drive are unclear about the basic road rules'. It could be suggested that to remain a safe and responsible driver and to lead the learning of other drivers, you should stay in touch with changes and updates in the road rules and road conditions. A teacher of science should also continually engage with thinking about how the discipline of science is evolving to address the challenges faced by society and, further, how the education associated with such evolution has responded to promote scientifically literate citizens. There is a need for all teachers of science to stay current with teaching approaches that best meet the learning needs of their students—particularly in relation to curriculum and assessment issues, effective use of technology in the classroom and other aspects of best practice. In terms of science education, this could also mean considering how the content and processes are shifting due to discoveries and breakthroughs: what is now considered content and processes of value and what is not, and why? So how might a teacher of science engage in such thinking while also monitoring the impact of new ideas on their practice?

PROFESSIONAL STANDARDS: PROVIDING A GUIDE

> Developing professional standards for teachers that can guide professional learning, practice and engagement facilitates the improvement of teacher quality and contributes positively to the public standing of the profession. The key elements of quality teaching are described in the Standards. They articulate what teachers are expected to know and be able to do at four career stages: Graduate, Proficient, Highly Accomplished and Lead. (AITSL, 2011, p. 1)

The Australian Institute for Teaching and School Leadership (AITSL) supports the federal, state and territory governments to promote excellence in the profession of teaching and school leadership, through the development of the AITSL Professional Standards. These standards make explicit the elements of high-quality teaching and attempt to define stages of career development and the dispositions that distinguish teachers at each stage of professional growth. The standards clearly lay out what you as a teacher need to know and what you need to pay attention to in order to perform in a professional manner. There are seven standards (Table 9.2) that have been organised around focus areas, each of which is explained for each of four stages of a teacher's career (Graduate, Proficient, Highly Accomplished and Lead). Visit www.aitsl.edu.au to view a complete version of the standards, their elaborations and the explanations for each of the four career stages.

Further support that may be used to inform your thinking about your professional learning needs and growth can be found within a set of professional standards produced by the Australian Science Teachers Association (ASTA). The National Professional Standards for Highly Accomplished Teachers of Science (ASTA, 2002) were established with the support of practising science teachers and look

Table 9.2 AITSL Professional Standards

Standards
Know students and how they learn.
Know the content and how to teach it.
Plan for and implement effective teaching and learning.
Create and maintain supportive and safe learning environments.
Assess, provide feedback and report on student learning.
Engage in professional learning.
Engage professionally with colleagues, parents/carers and the community.

to identify what it is that highly accomplished teachers of science know and are able to do (Table 9.3).

Table 9.3 ASTA Professional Standards

Highly accomplished teachers of science have a broad and current knowledge of science and science curricula, related to the nature of their teaching responsibilities.
Highly accomplished teachers of science have a broad and current knowledge of teaching, learning and assessment in science.
Highly accomplished teachers of science know their students well and understand the influence of cultural, developmental, gender and other contextual factors on their students' learning in science.
Highly accomplished teachers of science design coherent learning programs appropriate for their students' needs and interests.
Highly accomplished teachers of science create and maintain intellectually challenging, emotionally supportive and physically safe learning environments.
Highly accomplished teachers of science engage students in generating, constructing and testing scientific knowledge by collecting, analysing and evaluating evidence.
Highly accomplished teachers of science continually look for and implement ways to extend students' understanding of the major ideas of science.
Highly accomplished teachers of science develop in students the confidence and ability to use scientific knowledge and processes to make informed decisions.
Highly accomplished teachers of science use a wide variety of strategies, coherent with learning goals, to monitor and assess students' learning and provide effective feedback.
Highly accomplished teachers of science analyse, evaluate and refine their teaching practice to improve student learning.
Highly accomplished teachers of science work collegially, within their school community and wider professional communities to improve the quality and effectiveness of science education.

While both sets of standards act as frameworks against which teachers can consider their practice, very importantly each can be used to support teachers to pay attention to their own learning throughout their career. This can be done in ways that enable teachers to develop aspects of their personal professional practice through focused personal inquiry. In this way, professional learning and growth are contextually relevant and personally meaningful, rather than simply working to the next standard to meet external expectations. Further, the AITSL Professional Standards demand increasingly complex evidence of proficiency to successfully demonstrate a personal capacity to undertake the role and responsibilities of a professional teacher. This would suggest that monitoring and utilising the insights that this learning offers provides crucial evidence of proficiency. Just as evidence of learning growth in your students' needs to be strategically and systematically maintained, evidence of your own learning growth as a teacher needs to be maintained in such a way that it is a planned and organised endeavour.

Professional learning can be nurtured, explored, monitored and documented in many ways. Ideas include:

- working with a colleague to observe each other teach and then discussing what is observed, the intentions behind it and reflections on the teaching, and documenting these ideas in a learning log about your teaching
- keeping a reflective teaching diary (written, audio or video) to record interesting ideas, thoughts and reflections and sharing this with a critical friend to invite their feedback
- implementing something new in your classroom and then gauging students' responses in terms of engagement and learning while also considering your responses, then capturing this thinking in case writing to be shared with a wider audience to enable wider discussion of your experience.

These examples emphasise the value of teachers as professionals working together, sharing ideas about what matters for learning and teaching, and the importance of inviting students to feedback their experiences, providing a wider range of alternative perspectives. The implementation of the ideas outlined above and those discussed in detail as examples later rely heavily on a very important factor: the willingness of teachers to actively engage with their practice to develop their own professional learning, and maintaining this willingness throughout their career. Professional learning requires a personal commitment, so as a professional teacher you need to be willing to engage in reflective activities that interrogate your teaching practice, the learning of your students and your own learning. This demands a willingness to undertake the hard work of pushing your own ideas further and developing your practice in new ways. The outcome may be a rethink of what were strongly held beliefs and values about learning and teaching—in particular, rethinking your assumptions about learning and teaching science. Further, reimagining your practice requires you to give yourself permission to value and embrace the shifts in your thinking so that you are better placed to not only understand why you see a need to change but are able to articulate the reasoning and thinking behind these changes.

The level of engagement and commitment to professional learning required by teachers of science to achieve such shifts in their practice and thinking are important to consider; some teachers would view this as a necessary part of their role as professional science teachers and others would view it as far too difficult.

THE CONDITIONS THAT ACTIVELY POSITION TEACHERS AS PROFESSIONAL LEARNERS

> For teachers, learning occurs in many different aspects of practice, including their classrooms, their school communities, and professional development courses

> or workshops. It can occur in a brief hallway conversation with a colleague, or after school when counselling a troubled child. To understand teacher learning, we must study it within multiple contexts, taking into account both the individual teacher-learners and the social systems in which they are participants. (Borko, 2004, p. 4)

Borko's quote highlights that external professional development programs do not define the parameters of teacher learning. Every day, teachers talk together, sharing ideas and experiences; in doing so, they are actively working to understand and develop their practice. In this way, teachers as practitioners are continually immersed in the research endeavour of understanding the relationship between learning and teaching. As stated repeatedly, the key to continued professional growth is a willingness of teachers to continually interrogate and make sense of both their professional practice and their **professional knowledge**. It is reasonable to expect that their learning needs and interest will undoubtedly shift as they progress through their careers. What types of learning experiences enable teachers to develop personally meaningful and contextually relevant professional knowledge?

professional knowledge a combined knowledge of many aspects, including knowledge of curriculum, content, students and how they learn. All areas of teacher professional knowledge are interrelated and interdependent

social construction of professional knowledge emphasises the interactions between teachers and how they use language to construct a shared reality and understanding of teaching

Professional learning opportunities, which enable teachers to interact together and engage in focused professional conversations about their practice, appear to be particularly useful for enhancing teacher professional thinking and decision-making about their practice (Smith & Lindsay, 2016). When teachers work together to share the thinking that underpins their work, this allows them to engage in the **social construction of professional knowledge**. In critically reflective conversation, teachers maintain, modify and reconstruct a subjective reality, enabling some ideas to be shared. The social practices in which teachers engage therefore become crucial to the construction of meaningful understandings and shared knowledge about teaching.

As a result of these professional conversations, teachers reveal information that begins to identify the concerns to which they attend as they work to make sense of their experiences and construct their professional knowledge. Providing teachers with opportunities to work together and share their thinking through such professional conversations allows them to focus less on talking about effective teaching and instead talk more about learning and the important role of critical reflection in understanding and enhancing practice (Smith, 2017b).

Such learning experiences are important for the experienced teacher in terms of producing evidence that supports various aspects of professional standards. These experiences have also proven useful for pre-service teachers as they begin to prepare to undertake the role of a professional teacher. Tracking personal progress, completing personal reflections of teaching practice and compiling a professional portfolio that can be customised for a variety of readers enable teachers at all stages of their careers to consider teaching in a broader sense—particularly pre-service teachers when on professional placement:

> On reflection my objectives during this first placement were to avoid disaster, model my teaching on that observed in the school and meet the expectations of my supervisor as much as possible. Although at the time I would have stated that I was trying to create a rich learning environment for the students, I now

> realise that was not really my priority. I needed to feel comfortable and competent before I could really focus on teaching and then learning. It has taken me a while to realise that it is okay to get some things wrong at the beginning; not only is it inevitable but it is probably essential. I found that the experience of teaching in the classroom gives you a rich resource to criticise and improve on for next time as long as you take the time to reflect on it and are prepared to implement changes for next time. (Gwilym, 2015, p. 75)

This quote, taken from a case writing collection by pre-service teachers, captures the development of very insightful professional knowledge. Examining teaching placement through the strategy of case writing allows the pre-service teacher to reflect on and share aspects of practice specific to personal needs and interests. In doing so, the pre-service teacher articulates understandings and begins to develop new insights about teaching and learning. Strategies such as case writing encourage teachers at all stages of their careers to identify the moments in their practice when unexpected situations emerge, and then unpack or deconstruct those moments to determine why they found them curious, confounding, dissatisfying, rewarding or challenging.

Case writing, learning logs, digital stories and so on all provide examples of such learning opportunities, yet to be effective these strategies depend on the willingness of teachers to identify what they feel matters in their practice, rigorously interrogating critical moments—not merely sitting and thinking about what took place but actively digging deeper to identify why this moment mattered and how it contributed to or undermined the quality of the learning that was intended. All such strategies not only promote professional learning, but are also important because each provides evidence of such learning, documenting information from a professional careers stance, in particular personal growth and improved capabilities. Teachers need to actively engage in decisions concerning the direction, process and documentation of their own learning, while experimenting with new teaching procedures. In this way, teachers construct a **teacher knowledge of practice** that is connected and therefore contextually relevant to their teaching and learning practice. This information can then be articulated for multiple purposes and shared with multiple audiences.

teacher knowledge of practice a complex and deep knowledge about the processes and practices of teaching that guides pedagogy of effective teachers

While such conditions for learning are ultimately productive for personal learning, as Wilson and Berne (1999, p. 200) explain, this type of learning is 'hard work'. The following quote, again by a pre-service teacher, indicates that such learning is ongoing and constantly developing:

> Although I am still learning to teach I was able to draw on different types of knowledge, including repertoire of practices that effectively engaged the students. I recognise it is not only time that will build on my pedagogical content knowledge (Loughran, Berry & Mulhall, 2012) but also praxis, reflection with action. I continue to build on pedagogies that incorporate both cognitive and social constructivism, but more importantly unpack the science and make it relevant to their world not the laboratory. (McClelland, 2015, p. 63)

Teachers at all levels of their career are not always practised at, or provided with opportunities to undertake, such an investment in their own personal development; they may not be encouraged to actively question their own professional knowledge or to be personally suspect about their professional practice (Wilson & Berne, 1999). However, Mansfield (2016) theorises that teacher learning requires some challenge to equilibrium, and that as a result important personal learning emerges from times when teachers' existing assumptions are challenged.

Professional learning at all stages of a teaching career, from pre-service teacher to highly accomplished, must enable teachers to take ownership of their personal expertise, and develop and articulate strong personal purpose in their professional practice. However, creating the conditions that support teachers to develop their personal capacity to undertake such change requires everyone at all levels of teacher education to reconsider their roles and responsibilities. Everyone must work to effectively value and attend to the specific strengths and needs of the teachers they aim to support.

CONCLUSION

Your professional learning and your development through various career stages depends on your willingness to actively notice why you work in the ways you do. Once your confidence begins to build and you achieve a proficient standard of teaching, you will develop new perspectives that will invite further development, moving you towards competency and beyond to mastery. Expertise is so much more than knowledge acquisition; rather, it is a deep understanding of the application of such knowledge and the ability to execute aligned teaching action. As you work to build your expertise in teaching science, you will find effective ways to weave together your professional knowledge and experience so that your teaching nurtures quality learning. In doing so, you may also reach a level of competency that makes quality teaching look deceptively easy.

Building proficiency in any endeavour relies on a personal disposition that recognises rich opportunities for learning provided by challenge. Developing such a mindset is essential in science teaching because it will enable you to actively explore and attend to combinations of extremely complex demands, considerations and routines in ways that enhance your expertise. As you move through different career stages, utilise effective professional learning opportunities that enable you to notice and attend to your science teaching in different ways. Seek learning experiences that support you to interrogate your practice in ways that are personally meaningful, and above all take ownership of your own ongoing professional growth. Under these conditions, it is more likely that you will experience a deepening sense of purpose, and in turn develop your skills and your identity as a professional science educator.

DISCUSSION QUESTIONS

9.1 As a teacher of science in either a primary or secondary school setting, what strategies will you use to critically reflect on your teaching?

9.2 What strategies have you observed teachers using in schools to monitor their professional learning in ways that are structured and organised?

9.3 This chapter includes information about the AITSL Professional Standards. How might you use these standards to identify a focus for your professional learning in relation to science teaching?

9.4 How will you document and develop an evidence-based approach to researching your own practice?

9.5 Professional conversations have an important role to play in professional growth. How might you foster opportunities for discussion as part of your own practice? What steps or initiatives might you adopt?

FURTHER READING

Corrigan, D., Dillon, J. & Gunstone, R. (2011). *The Professional Knowledge Base of Science Teaching*. Dordrecht: Springer.

REFERENCES

AITSL (2011). *Australian Professional Standards for Teachers.* Canberra: Education Services Australia as the legal entity for the Ministerial Council for Education, Early Childhood Development and Youth Affairs (MCEECDYA). Retrieved from www.aitsl.edu.au/docs/default-source/apst-resources/australian_professional_standard_for_teachers_final.pdf.

ASTA (2002). *National Professional Standards for Highly Accomplished Teachers of Science*. Melbourne: Australian Science Teachers Association and Monash University.

Bartholomew, H., Osborne, J. & Ratcliffe, M. (2004). 'Teaching students "ideas-about-science": Five dimensions of effective practice'. *Science Education*, 88(5), 655–82.

Borko, H. (2004). 'Professional development and teacher learning: Mapping the terrain'. *Educational Researcher*, 33(8), 3–15.

Brookfield, S.D. (1995). *Becoming a Critically Reflective Teacher.* San Francisco: Jossey-Bass.

Clandinin, D.J. (1985). *Classroom Practice: Teacher Images in Action.* London: Falmer Press.

Corrigan, D. & Gunstone, R. (2011). 'An approach to elaborating aspects of a knowledge base for expert science teaching'. In D. Corrigan, J. Dillon & R. Gunstone (eds), *The Professional Knowledge Base of Science Teaching.* Dordrecht: Springer, pp. 83–106.

Day, C., Sammons, P., Stobart, G., Kington, A. & Gu, Q. (2007). 'Teachers' professional life phases: A research informed view of career long effectiveness'. In C. Day, P. Sammons, G. Stobart, A. Kington & Q. Gu, *Teachers Matter: Connecting Lives, Work and Effectiveness*. Maidenhead: Open University Press/ McGraw Hill Educational.

Fuller, R. (2013). 'Teaching isn't rocket science. It's harder'. *School Reform and Classroom Practice*. Retrieved from: https://larrycuban.wordpress.com/2013/12/25/teaching-isnt-rocket-science-its-harder-ryan-fuller.

Gwilym, S. (2015). 'Small steps that start a long journey'. In S. Keast, L. Rutherford, S. Gwilym, A. Hooper, K. Marangio & D. Panizzon (eds), *Learning to Teach Science: Stories of Practice by Pre-service Teachers.* Melbourne: Monash Print Services.

Hargreaves, A. (1994). *Changing Teachers, Changing Times: Teachers' Work and Culture in the Postmodern Age*. London: Teachers College Press.

Huberman, A. (1989). 'On teachers' careers: Once over lightly, with a broad brush'. *International Journal of Educational Research*, 13, 347–61.
—— (1993). *The Lives of Teachers*. London: Teachers College Press.
Loughran, J.J. (2010). *What Expert Teachers Do: Teachers' Professional Knowledge of Classroom Practice.* London: Routledge.
Loughran, J., Berry, A. & Mulhall, P. (2012). 'Portraying PCK'. In *Understanding and Developing Science Teachers' Pedagogical Content Knowledge.* Rotterdam: Sense, pp. 15–23.
Mansfield, J. (2016). 'Exploring the nature of pedagogical equilibrium'. Unpublished PhD thesis, Monash University.
McClelland, K. (2015). 'Developing complexity in teaching'. In S. Keast, L. Rutherford, S. Gwilym, A. Hooper, K. Marangio & D. Panizzon (eds), *Learning to Teach Science: Stories of Practice by Pre-service Teachers.* Melbourne: Monash Print Services.
Opfer, V. & Pedder, D. (2011). 'Conceptualizing teacher professional learning'. *Review of Educational Research,* 81(3), 376–407.
Simos, M. (2011). 'Parents unsure of road rules'. *The Advertiser* [Adelaide], 18 February.
Smith, K. (2017a). *Teachers as Self-Directed Learners: Active Positioning Through Professional Learning.* Dordrecht: Springer.
—— (2017b). 'Conditions for learning: Building and sharing teacher professional knowledge'. In K. Smith & J. Loughran (eds), *Quality Learning: Teachers Changing Their Practice* (Vol. 22, pp. 29–38). Rotterdam: Sense.
Smith, K. & Lindsay, S. (2016). 'Building future directions for teacher learning in science education'. *Research in Science Education*, 46(2), 243–61.
Speakman, J. (2012). 'What are we really trying to teach?' In S. Keast & J. Loughran (eds), *Exploring Scientific Literacy Through Cases of Teaching and Learning*. Melbourne: Monash Print Services, pp. 18–19.
Timperley, H. (2011). *Realizing the Power of Professional Learning.* London: McGraw-Hill.
Tripp, D. (1993). *Critical Incidents in Teaching: Developing Professional Judgement*. London: Routledge.
Wilson, S.M. & Berne, J. (1999). 'Teacher learning and the acquisition of professional knowledge: An examination of research on contemporary professional development'. *Review of Research in Education*, 24, 173–209.

CHAPTER 10

SCIENCE EDUCATION IN INFORMAL SETTINGS

ALAN REID *and* BEN LIU

LEARNING OUTCOMES

After reading this chapter, you should be able to:

- articulate and evaluate your beliefs about where science education happens, and what conventional to innovative practice looks like in a range of settings
- describe and explain why informal settings matter for science education in terms of helping realise what is possible and desirable in science learning and teaching
- decide and justify what should be prioritised in ensuring that an appropriate range of places is utilised in purposeful and authentic science education
- recognise and address some of the barriers to learning about science in informal settings, knowing what to consider when examining these and related matters further.

This chapter starts with a case study of a successful partnership between a primary school and leading science institution on a science-related topic. Case Study 10.1 illustrates how collaboration in a program happens in ways that develop and communicate the value of a more authentic science education (Corrigan, Dillon & Gunstone, 2007) through the lens of education for conservation in an outreach program from a zoo.

CASE STUDY 10.1: Real-life science literacy—and action

Great Ocean West Primary School's partnership with Zoos Victoria began in 2013 and is based around the zoo's marine entanglement program, Seal the Loop. The Seal the Loop initiative aims to do the following:

1. Reduce marine wildlife entanglement rates:
   - Fishing line entanglements result in the death of an estimated 1400 seals every year in Australia.
   - Nearly 30 marine animals including seabirds, turtles, whales, dolphins and sharks are listed by the Australian government as being 'at risk' due to the key threatening process of injury and fatality caused by ingestion of, or entanglement in, harmful marine debris.
   - Three seal species breed in Australia (Australian sea lions, Australian fur seals and New Zealand fur seals), and all three have entanglement rates that are among the highest reported for any species of seal in the world.
2. Raise awareness of the threats that plastics pose to marine wildlife, using educational programs and interpretive static displays at the zoo.

3 Encourage onsite and offsite recycling, consequently reducing the amount of waste generated at Melbourne Zoo and at its partners in the program.

Seal the Loop resonated with the teachers and students at Great Ocean West Primary School, due to their close proximity to both the coast and a population of Australian fur seals. The school was one of the first to embrace Zoos Victoria's Seal the Loop program while visiting Melbourne Zoo, and worked with the zoo education team to find out how it could get involved further.

In the first year, senior students engaged younger classes in an action day event that involved a large clean-up of their local beach and a number of educational activities, in collaboration with Zoos Victoria.

In the second year, again working alongside Zoos Victoria, students expanded the event to host a community day. They

Figure 10.1 Education for Conservation: Seal the Loop Action Day—schools undertaking community conservation campaigns in their local community
Source: Zoos Victoria.

presented on the issue of marine entanglement to other local schools and community groups, and were involved in presenting marine debris data collected by students to local fishing and community groups, highlighting the increasing threat of fishing line entanglement to local wildlife.

Working closely with Zoos Victoria to further develop the program in its second year enabled the students at the school to become mentors to other schools and community groups in their third year. They ended up pioneering an approach for school and community involvement that Zoos Victoria was then able to share with other schools across the state.

Through its collaboration with Zoos Victoria and adopting a more purposeful approach to primary science education, the school received the award for Outstanding Curriculum Innovation at the Victorian Education Excellence Awards. The following quote from the school's science specialist demonstrates how collaboration with Zoos Victoria has enhanced science education across the school:

> This partnership has been about working together but also getting something out of it for both parties. For Zoos Victoria, it is about spreading the conservation messages while for us, the science partnership has become a point of difference as a school and engaged our kids is real-life problems that need solving. Because we are literally one sand dune away from the ocean, the real-life problems are in students' own backyards, giving them even more reasons for getting involved.

As we show throughout this chapter, Zoos Victoria offers an authentic site for developing and examining science education, given the ways in which it supports education and provides opportunities for schools and communities to work in partnership in an informal setting for science education to foster learning outside of school:

> Learning occurs when the meaning people make of their experiences results in change, which may be cognitive, affective, psychomotor, social, or cultural and though not necessarily visible, can be inferred through behavioral change in the learner. Learning is personalized, contextualized, and cumulative . . . These characteristics have particular salience when considering learning outside of school, because out-of-school contexts usually involve a degree of volition, so individuals' unique backgrounds and experiences and the choices they make about the nature of their activities can bring about very different learning outcomes. The nature of these outcomes is also influenced by the sociocultural context of those activities . . . Many out-of-school science related activities, such as museum visits or excursions, are of short duration, so the impact of those experiences may come later; thus the cumulative nature of learning becomes an important factor when evaluating the contribution of out-of-school experiences to learning in science. (Rennie, 2014, pp. 120–1)

Case Study 10.1 illustrates what is possible when the purpose that arises from collaborations is based around three key characteristics:

- schools that partner with leading science institutions, whose missions support each other through addressing scientific issues embedded in society (e.g. preserving biodiversity)
- schools and institutions that are co-located within a localised community context

- schools and institutions that regard education as a vehicle for addressing matters of public concern (e.g. conservation, scientific literacy, public understanding of science).

The chapter is co-authored by a zoos education researcher and a zoo educator, to explore a series of themes and insights arising from reflections on the work of Zoos Victoria, and informal settings for science education more broadly. We encourage you to consider how representative this extended case study is of science institutions in general, and informal settings used for science education near and far, as you pursue the learning objectives for this chapter.

KEY CONSIDERATIONS

Before we profile and evaluate the work of Zoos Victoria in more detail, we flag some of the key considerations from the research literature on science education in relation to the place and roles of zoos in education more generally:

> A permanent cultural institution which owns and maintains captive wild animals that represent more than a token collection and, under the direction of a professional staff, provides its collection with appropriate care and exhibits them in an aesthetic manner to the public on a regularly scheduled basis. They shall further be defined as having as their primary business the exhibition, conservation and preservation of the earth's fauna in an educational and scientific manner. (AZA, 2017; see also EAZA, 2008)

The literature invites questions about and reflections on how to get the most from doing science education beyond the classroom when the focus is on using local zoo-based education as a site for informal and non-formal science education. The notion of informal science education reminds us that the learning and teaching of science happens not only in the formal confines of schools, colleges or universities, but can happen through a range of experiences and settings that are being designed, implemented and assessed by a community of dedicated, trained practitioners. Spaces and places for informal science education are diverse, and include zoos, aquariums, gardens, parks, education and nature centres, libraries, galleries, museums and so on. These settings may also be used for non-formal science education, with the key distinction being that the non-formal learning happens when learners themselves decide to engage in science learning and teaching, rather than the learning being part of a formal or informal education program. This relates to:

- learning and teaching (Adams, Gupta & DeFelice, 2012)
- what is typically assumed (Roe, McConney & Mansfield, 2014b) and experienced (Ballantyne et al., 2007) as best practice in zoo-based education, and how these square with critical perspectives on the public (Carr & Cohen, 2011) and commercial (Westley & Vredenburg, 1996) dimensions of zoos
- whether it is possible to draw a hard line between education and 'info-tainment' (Lucas, 1991) in learning about science at zoos (Maynard, 2017), and why these are often synergistic activities (Packer & Ballantyne, 2004)
- the range of intentions that teachers and learners may have in visiting zoos (Roe & McConney, 2015), including the effect of past experiences on what might be learnt or taught there (Yocco et al., 2015) and experienced during encounters with 'curated nature' (Grazian, 2012) by both students and teachers
- how to address the connections (and disconnects) in teacher, researcher and learner views of science education with a particular purpose in mind—for example, using interactive exhibits

at zoos (Rennie & McClafferty, 1995) or animal viewing and encounters to connect people with nature (Bruni, Fraser & Schultz, 2008)
- the roles of pre-visit knowledge development, and on-site and follow-up activities (Bueddefeld & Van Winkle, 2018) in underpinning (as much as problematising) science learning in zoo settings—for example, to foster knowledge of animal biology (Dove & Byrne, 2014) or further species-specific (Ballouard et al., 2012) or habitat (Pearson, Dorrian & Litchfield, 2013) understandings, and how such imperatives might be challenged (Neves & Monteiro, 2014)—including in relation to conservation-focused imperatives (Fennell, 2013).

As set out in the learning objectives for the chapter—and as a contribution to the book as a whole—we invite you to revisit these topics in relation to both your assumptions and experiences of where science education typically takes place, and why the spaces for science education matter to what is possible in science learning and teaching:

> Victorian students love their local species. Education programs pre-Education for Conservation have traditionally focused on iconic endangered species (for example, gorillas, tigers and orangutans). But with the move to align Education for Conservation to the conservation work of Zoos Victoria, this has meant a larger focus on 21 locally endangered species that Zoos Victoria is working to protect. Shifting the focus from the world's iconic endangered species to locally endangered species has been successful, with more than 50 per cent of the schools undertaking conservation action, doing so for one of the 21 locally endangered species. (Ben Liu, Zoos Victoria)

Linking both is a third consideration threading throughout this chapter: how we might step back to examine where we tend to be when we immerse ourselves and others in science education. Why? Because all three considerations invite us to reflect on the value of pursuing authentic questions and experiences for a purposeful and innovative science education (Jenkins, 1994)—especially by way of making good use of informal science learning environments.

WHERE TO BEGIN?

The American humourist Evan Esar once quipped that zoos were excellent places to study the habits of human beings. In this chapter, we rework this left-handed compliment, inviting you to first evaluate the 'habits' and 'habitats' of science education with which you may have grown up then, by way of the various trips to the zoo we make in the following pages, consider what we can learn from zoo-based education to foster better learning and teaching of science across the year levels.

We begin our journey from the familiar idea of a landscape and its less familiar partner notions, the soundscape, smellscape, cityscape, technoscape and ethnoscape. The suffix 'scape' highlights the ways in which we can be influenced by a particular feature of our surroundings even as we recognise certain boundaries and flows of what is, or becomes, present or absent in that 'scape'. For example, our senses (as much as our common sense) might help us distinguish between biotic and abiotic elements (landscape), silences and noises (soundscape), scents and odours (smellscape), built surfaces and their configurations (cityscape), technological enhancements or obstacles (technoscape), and the visibility or invisibility of dominant to marginal cultures (ethnoscape).

In ecology and biology, abiotic factors refer to non-living physical and chemical elements in an environment, that affect an ecosystem. Abiotic resources are usually obtained from the lithosphere, atmosphere and hydrosphere. Examples of abiotic factors are water, air, soil, light, gases, pH levels, temperature and minerals. This is in contrast to biotic factors, those living or once-living organisms in an environment that affect an ecosystem. These are obtained from the biosphere and are capable of reproduction. Examples of biotic factors are animals, birds, plants, fungi, bacteria and other similar organisms. In this, the places for any of these 'scapes' to occur might have required some design to foster particular experiences, appreciations and outcomes, including those that afford (or avoid!) particular teaching and learning opportunities (Figures 10.2 and 10.3). Equally, these 'scapes' can become a set of (perhaps largely

Figure 10.2 Education Resource Centre Library with giraffe pelt

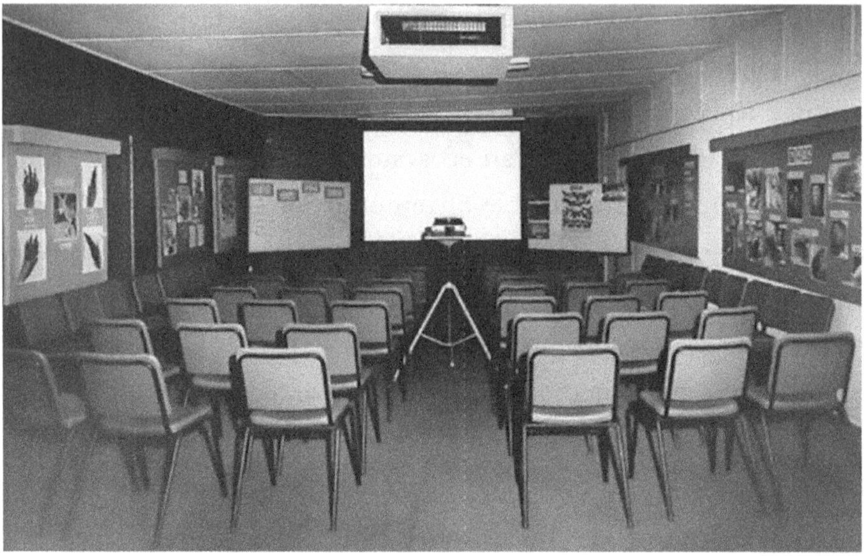

Figure 10.3 Zoo School: Inside the first Zoo School classroom at Melbourne Zoo
Source: Zoos Victoria.

unreflective) circumstances for people's lives and learning, because of what others do or expect a certain place to be for them or you—for example, during a visit to the school laboratory or zoo (Figure 10.4) (Powell & Bullock, 2014).

Putting this differently, while some people may pay attention directly or indirectly to certain features of their surroundings in their day-to-day lives, more often than not our focus is elsewhere—for good or ill—including on what we consider ensures or disrupts the best spaces and places of science education (Schultz & Joordens, 2014; Wagoner & Jensen, 2010). So as we leave our point of departure, you might ask yourself and your peers what is taken for granted as both the obvious and overlooked features of the spaces and places of science education—their sounds, noises, smells, activities, cultures and subcultures—as well as the advantages and disadvantages of some of the mainstays of and alternatives to these.

For example, consider the familiar four walls of a science

Figure 10.4 Zoo School: The first Zoo School classroom at Melbourne Zoo, now The World of Frogs
Source: Zoos Victoria.

laboratory and what takes place within the labs you have known, or that your students have experienced. Is there a particular set of smells you associate with that place? Are there certain kinds of noises, dress codes, instrumentations, work surfaces, subcultures and so on that set this 'scape' apart from other places in schools, or even bring other kinds of 'scapes' into the school—like those of the world of work or particular forms of scientific enterprise?

Next, consider which species, apart from humans (or 'laboratory animals'), you expect to encounter as part of science education, and how this might occur (Figures 10.5, 10.6, 10.7). Again, are there certain kinds of noises or smells or tactile interactions that tend to become a feature of this 'scape' for science education, given their particular presence or absence? To what extent do these places and their occupants influence what might be regarded as appropriate and fitting for learning and teaching about science?

Figure 10.5 Teaching in the Zoo School: Zoo educator with students and animal encounter
Source: Zoos Victoria.

Figure 10.6 Student feeds a tawny frogmouth, Healesville Sanctuary
Source: Zoos Victoria.

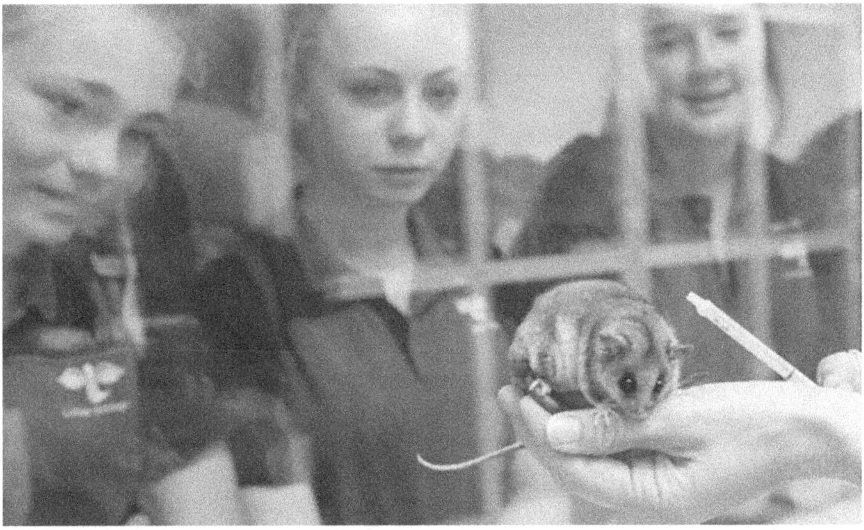

Figure 10.7 Education for Conservation: Conservation focused animal encounters—students meet a mountain pygmy possum
Source: Zoos Victoria.

Finally, what do you remember about such sensorial and cultural dimensions of the 'scapes' of science education, and the encounters with others (including other species) that you have had there? Have they proven to be, for example, significant or largely insignificant as life experiences for your learning of science, and continuing interest in it (Figures 10.8 and 10.9)? For example, was fieldwork more or less evocative to you, and why? How did that square with what you learnt from textbooks or screens? And finally, what else do you desire from such 'scapes' for science education: could the places or spaces for science learning and teaching become something different—even more meaningful or challenging—by changing them or going somewhere else?

A FRAMEWORK FOR CONSIDERING SCIENCE-SCAPES

Before we provide some additional thumbnail sketches of science-related learning activities drawn from Zoos Victoria, we set out a

Figure 10.8 Zoo School—towards less formal interactions?
Source: Zoos Victoria.

Figure 10.9 Secondary students inspired by a lion keeper talk, Werribee Open Range Zoo
Source: Zoos Victoria.

framework for critically considering the contribution these 'scapes' might make to science learning and teaching, before discussing an additional scape, the science-scape. We also invite you to do some lateral thinking about this framework and the examples from the zoo. This is because, while science education is most typically associated with the formal setting of the school, such as indoor and outdoor classrooms or laboratory spaces, these may only suit particular forms of collaboration, innovation and active learning among students and teachers (Andersen, 2003). In other words, keeping the notion of a 'scape' to the fore reminds us that we might also *choose* a particular learning environment for science education, rather than accepting the default option provided. In so doing, we might be able to ensure we utilise that which is most suited and stimulating for the focus and intended outcomes of any particular science learning and teaching, and if not, raise the critical question of why not?

For Heimlich, Adams and Stern (2017, p. 117), choices of 'scape' are linked to a way of categorising the type of program, level of choice for participants, typical goals of participants and typical goals of providers. To develop a critical perspective of science education in general, we could ask the anthropologist's reflexive question: what would be different if the 'scape' for the visit was the usual spaces used for school science, such as the school laboratory, classroom or school grounds?

SCIENCE-SCAPES

Another consideration arising from Table 10.1 harks back to the celebrated twentieth-century media theorist Marshall McLuhan (1964). For McLuhan, if the *medium* is also in some sense the *message,* where and when sensory and cognitive experiences happen during science education will matter immensely to realising and evaluating its outcomes. So if we now add a further 'scape' to our list, the *science-scape*, we can recognise that *where* science

Table 10.1 Type of education program according to participant choice and goals and provider goals

Type of program	Choice of participation	Typical goals of participants	Typical goals of providers
School-related field trips or related programs	None	Entertainment Academic achievement Content knowledge	Providing content knowledge Influencing conservation behaviour Teaching critical thinking
Casual visit to a community institution (e.g. a zoo)	Variable	Social experience Educational	Specific content Environmental conservation or protection attitudes and behaviours
Specific programs (e.g. an extension or outreach presentation or stewardship activity)	High	Knowledge of topical or contextual information Skills development Social experience Addressing a specific problem	Specific context or skill development Environmental conservation or protection behaviours
Recreational programs	Variable	Exercise Fun Social experience	Health Environmental awareness Conservation behaviours

education happens (or doesn't happen) very much matters to what can be experienced, taught and learnt, and thus to what is appreciated or expected within particular spaces for science education. A science-scape refers to those spaces and places that deliberately foreground the creation, representation and/or exchange of scientific knowledge—for example, a school laboratory, a science learning centre, an R&D park, a botanic garden learning zone, an exhibition, or a public event on a scientific advance or controversy. In addition, a science-scape may well emerge in (1) film and broadcast media, (2) cyber-learning and gaming, and (3) youth, community, and out-of-school time programs; while the first two of these might also afford examples of techno-scapes—for example, through the use of augmented reality or immersive visualisation techniques.

Informal spaces for science education are diverse, obvious and perhaps even surprising. For example, in Victoria, Australia, some informal science institutions include Scienceworks, the Victorian Space Science Education Centre (VSSEC), the CSIRO Discovery Centre, Zoos Victoria, Parks Victoria, the Botanic Gardens and the Centre for Education and Research in Environmental Strategies (CERES). Their educational programs and environments are often designed and developed in particular ways to achieve particular educational, social-behavioural, environmental or aesthetic outcomes that aren't easy to achieve in the formal setting of the classroom (Fraser & Sickler, 2008). In this chapter, we consistently use zoos as an example of informal spaces, but reader, do consider these other types and examples when engaging with this topic (e.g. Griffith & Symington, 1997; Vernon & Boyle, 2008).

At a zoo, key reasons for experiencing or possibly interacting with wild or domesticated animals may reflect a variety of purposes (Dohn, 2013):

- to observe behaviours as individuals or in groups
- to interact with experts in real time to develop vivid understandings of conservation activities and human-animal relationships

- to examine moral and ethical dimensions or dilemmas related to the science and practice of having animals in captivity.

Experiencing this brings the sensory and somatic dimensions to a science-scape into sharp relief, by considering the impressions and thinking that are awakened via such encounters at zoos (including what is evoked in the figures above). Are these likely to be different to those of the traditional classroom and textbook exercise, and what might these differences mean for authentic and purposeful science education?

To develop this point further, informal science institutions are not new as sites for education, and have long been a part of a suite of possible educative experiences available to those more typically confined in schools (Rennie & McClafferty, 1996). For example, the scale of the visitations to zoos should not go unnoticed. Most often taking the form of a traditional field trip or excursion, each year accredited members of the Association of Zoos and Aquariums (AZA) reach over 180 million visitors across America, including 51 million students (AZA, 2018). Across the Australasian region, this figure is approximately 17 million visitors, including over 600,000 students annually. However, as Rennie (2014) notes, a key challenge is whether these visits are cumulatively beneficial to student learning, not just as an event or entertainment in and of themselves—that is, science educators will need to consider both the suite of experiences and the scope and sequence of learning science beyond the classroom in these alternative science-scapes. We would also add a further challenge: whether these contradict, dovetail with or extend what is learnt within the traditional science-scape(s) of the school—in other words, what values, and whose, students and teachers are engaging with in relation to education and conservation in these places for science learning (Pearson et al., 2014).

BEYOND THE SCHOOL TRIP

We now shift our focus to illustrating that much more is possible for authentic science education than mere field trips—important and memorable though these can be. This is because, for some aspects of science education—particularly cross-curricular or interdisciplinary work—learning and teaching in informal settings can make more sense for achieving particular educational outcomes than staying in the most familiar or everyday setting for science education (Dillon, 2016). So, rather than have a zoo treated as a surrogate classroom for science education, we now invite you to consider what science education might become when it goes beyond its usual classroom settings (Wünschmann et al., 2017). Case Study 10.1 may be considered an example of what can happen when such invitations are taken up positively: the science learning has been extended beyond both the school classroom and the zoo as a surrogate classroom, but has also been deepened and reconfigured well beyond the field trip.

To unpack what we mean by this, we must recognise that while the learning that can happen in places of informal learning such as science institutions may replicate that in the formal setting of the school in broad terms—for example, creating a 'zoo school' (Hopkins, 1991), it can also be organised in diverse ways with quite different ends in view. On one hand, the zoo may offer active and guided inquiries undertaken by students with zoo educators, often with the assistance of their usual teachers (Matiasek & Luebke, 2014) (Figure 10.10). On the other, these means and ends may try to harness what might be learnt through 'free choice' or incidentally during an out-of-school visit, such as during visits to learning zones or through learning quests as part of an excursion to the zoo. Traducing both, they might be organised very differently because the responsibility for the science learning and teaching is recognised as shared with or given to other educators (Heimlich & Falk, 2009), such as members—junior to senior—of the zoo education team.

Figure 10.10 Secondary students planning the locations of their inquiries, Werribee Open Range Zoo
Source: Zoos Victoria.

To illustrate, most of the aforementioned institutional examples have education programs that try to move away from the habits of the past, associated as these often were with the traditional static display of collections or artefacts, encounters with other species or accumulating content knowledge (Dierking et al., 2002). In other words, rather than invoke the equivalent of dioramas as popularised in Hollywood movies, science educators (e.g. Rennie, 2014, p. 125) argue that teachers and learners should experience something both familiar and different when doing science education in these out-of-school places, including in the company of—perhaps some quite unfamiliar, scary or differently formed—others.

In our next section, we provide a brief context and overview for understanding what happens when this challenge is taken up at Zoos Victoria, before raising some questions about how we can evaluate what is it appropriate and worthwhile as part of a well-rounded and meaningful science education, within and beyond the walls of a school setting.

WE'RE GOING TO THE ZOO—HOW ABOUT YOU?

Using the specific examples of zoos and thinking about the frameworks and ideas provided above, teachers of science need to consider seriously what is it they wish to achieve in collaborating with an organisation such as a zoo. If it is truly a collaboration, how do they consider any such activity from the perspective of their students (what is to be learnt?), their partners (what is in it for the zoo?) and for themselves (what learning opportunities can be provided that cannot be provided in school or in such authentic ways?).

Case Study 10.2 outlines some of the key features of the approach taken at Zoos Victoria that illustrates the 'specific program' type in Table 10.1. A key rationale for combining the two streams—biological and educational—at Zoos Victoria is noted by Roe, McConney and Mansfield (2014a, p. 259). They suggest that, with their significant reach, environmental organisations such as zoos, museums and parks are 'a valuable resource and ideally positioned . . . for reconnecting an increasingly urbanised public with the natural world'.

Put differently, for a large majority of an increasingly urbanised population, conservation and informal science organisations and their sites are among the few avenues available for engagement and education that foster a broad-based public understanding of science—for example, about the biotic and abiotic environment. Even if activists or critics object to using zoos as appropriate or necessary science-scapes for learners and teachers to engage with, it remains the case that it is difficult to raise political awareness or even the sheer will necessary to address the conservation challenges currently faced by the globe (Amir & Abraham, 2010) without a diverse range of public science-scapes (see also Levinson, 2006, for a framework of how educators might approach controversial issues in science education).

CASE STUDY 10.2: Education for Conservation

In 2015, Zoos Victoria launched its Education for Conservation program across its three properties: Melbourne Zoo, Healesville Sanctuary and Werribee Open Range Zoo. Education for Conservation was designed to develop independent learners who would demonstrate positive environmental values and take action to protect wildlife and wild places. Implementing the program has required the redesign of full-day informal learning experiences on site at Zoos Victoria properties, as well as the addition of formal learning opportunities off site at schools.

By way of background, over 40 years ago Zoos Victoria designed an education model that consisted of up to twenty school classes rotating through a specially designated area of the zoo, affectionately known as the Zoo School. A key feature of this approach was the delivery of 45-minute scripted lessons focusing on key science curriculum areas. While this

Figure 10.11 A zoo-based conservation organisation: Zoos Victoria Fighting Extinction

Source: Zoos Victoria.

model positioned Zoos Victoria as a leader of zoo-based education at the time, and served Zoos Victoria well for many years, it did not address the zoo's wildlife conservation mission. Further, the scripted lesson approach no longer represented best practice in education. To address these shortcomings, and to halt recent declines in student visitation, Zoos Victoria redesigned its education model in 2014. The new model ensured that all programs would educate and enable students to undertake actions that would positively impact wildlife locally and internationally.

Prior to the implementation of Education for Conservation, Zoos Victoria's education programs traditionally focused on iconic endangered species (e.g. gorillas, tigers and orangutans). The educational offerings at Zoos Victoria now align directly with the wildlife conservation work of Zoos Victoria's wildlife science team, bringing the biological and educational streams of the zoo together under the banner of Education for Conservation.

The Education for Conservation model is driven by five key themes: conservation, pedagogy, unique opportunities, curriculum and capacity. For conservation, all programs engage schools with one or more of Zoos Victoria's conservation campaigns. For pedagogy, on-site programs are challenge-based and promote the values of inquiry-based learning. For unique opportunities, schools explore the whole zoo property and engage in at least six animal encounters that are unique to Zoos Victoria. For curriculum, programs link to learning outcomes specified in the Victorian Curriculum and promote the application of higher-level learning (e.g. analysis and evaluation of knowledge). Finally, for capacity, programs are designed to broaden Zoos Victoria's reach, including by increasing visitation and offering professional learning opportunities for educators.

> The best part of our students' experience was being able to ask questions and encounter a range of animals as they wandered around Healesville Sanctuary. (R. Waterworth, Derinya Primary School, Victoria, Australia)

These themes are brought together in activities that foster transformative student experiences, both on site at Zoos Victoria properties and off site through opportunities to join the Fighting Extinction Schools community.

More information about the case study is available through the links in the Further Reading section.

Figure 10.12 Education for Conservation: Out of the classroom exploring the zoo in small groups
Source: Zoos Victoria.

Notwithstanding these concerns, we note that every year, hundreds of schools visit Zoos Victoria properties as part of their learning and teaching program and, of these schools, many go further, collaborating with the zoo to rally behind the purpose of wildlife conservation. Through these collaborations, schools have been able to take deep dives into the biological science behind various conservation campaigns and drive incredible support for actions that support wildlife conservation. While the links in the

Further Reading section go into this in more detail, what this has meant for science education within these Victorian schools is that the learning and teaching of science has been given a very distinct and authentic purpose as it invites some action to be taken by participants.

In Case Study 10.1 (above) and Case Study 10.3, we sketch two examples of Victorian schools successfully collaborating with Zoos Victoria to achieve significant wildlife conservation goals through science- and action-focused learning experiences. Zoos Victoria's mission is centred around wildlife conservation, and it is particularly interested in encouraging behaviours that protect local wildlife as outlined by its stated goal of making sure that 'No Victorian terrestrial, vertebrate species goes extinct—not on our watch!'

CASE STUDY 10.3: Designing science learning that engages learners and teachers – *Survival*

For the Year 5 students at Eastern Meadows Primary School, what started as a small idea in 2014 grew into a three-year science education project that is now making its way into classrooms across Victoria. The journey began with students wanting to help one of Australia's most endangered frogs, the southern corroboree frog, after visiting Zoos Victoria's Healesville Sanctuary campus on a regular excursion. Through additional research, students came to understand that many Australian frogs were endangered and, as a school community, they decided they wanted to do something.

Drawing upon their teacher's previous experience as a game designer, the class decided to design and develop a cooperative card game called *Survival*. The idea of the game was to draw upon relevant and current conservation science to return stable populations of each frog species to the wild by researching, managing threats and taking positive actions. It was decided that the game must be cooperative (because organisations must work together in real life to protect our frog species); difficult to win (because conserving a species is something that is extremely difficult to do in real life); and educational (after playing, people should have an understanding of the life cycles of frogs, the threats frogs face, including chytridiomycosis (an amphibian fungus disease), and some of the things that can be done to conserve them).

The students and their teacher were so committed to the production of this game that it was decided to keep the class together for 2015 so they could continue testing and developing the game. Over the course of the next year, the students met with Zoos Victoria and other amphibian experts to obtain donated images and information, which helped to refine the game. For their efforts, the students went on to win the Young Advocate of the Year award at the Zoos Victoria Hall of Fame.

The students, who graduated to secondary school in 2016, are still working together to professionally produce *Survival*, in collaboration with Streets Ice Cream. They have also continued to work with Zoos Victoria to share copies of the game with Victorian schools.

WHAT ARE WE LEARNING?

As invited earlier, while we highlight some of the key outcomes of implementing the new Education for Conservation model across Zoos Victoria below, this is where some lateral thinking to pre-service education is in order. For example, we often overlook the fact that students from primary schools go on to secondary schools, and we might successfully extend and rework the science learning they had during an earlier phase by revisiting the zoo for these purposesor to take such learning into new areas. Equally, we might consider whether the learning at the zoo builds, blurs or corrodes the science education that is offered in formal or other informal settings.

Thus part of the lateral thinking we invite you to do includes reflecting on the parameters, direction and expectations of a career in science education, such as (1) where science education takes place; (2) what makes a particular science-scape meaningful and authentic and whether there are viable alternatives to the status quo—for example, through partnering with a science institution; and (3) what to prioritise to motivate learners and educators as well as ensure a high-quality science education for all (Fensham, 1985).

For instance, some of the outcomes of research into Education for Conservation approach are also equally applicable to science learning more broadly in light of these three considerations. These include the following (Liu & Reid, 2017):

- *Increased staff motivation.* Staff motivation has increased and stabilised as elements of the new model have become more embedded within the strategy and operations of the zoo, and as Zoos Victoria has captured student conservation stories from the new model. Many educators now see their work as aligned more closely with their original intentions in becoming a conservation educator in the first place—for example, bringing science to life with learners. They are also critically reflecting on their work using an evaluation, research and development strategy that is shared and supported by the zoo's leadership.
- *Education positioned as an investment in conservation.* The internal positioning of education as an investment in conservation has been one of the greatest outcomes of Education for Conservation for the organisation, and as a model for other

zoos and conservation organisations to consider. The strategic shift to align all education activities with the wildlife conservation work of Zoos Victoria has meant that capital outlay into education activities is now considered an investment in conservation, and can be evaluated in terms of a 'return on investment' model. As a result of this repositioning, educators are able to focus on developing student and teacher programs that equally achieve conservation outcomes and increase school visitation.

- *The creation of a conservation school community.* By its nature, Education for Conservation has led to more schools undertaking actions that support the conservation priorities of Zoos Victoria. This has led to the development of a network of like-minded schools across the state known as the Fighting Extinction Schools community. More than 50 per cent of the schools undertaking conservation action are doing so for one of the 21 locally endangered species that Zoos Victoria is working to protect.
- *Increased student visitation.* Increasing student visitation was a priority for the Education for Conservation model. In the initial eighteen months of implementation, Zoos Victoria saw a modest increase in visitation of approximately 5 per cent.
- *Changes in student knowledge, attitudes and behaviours.* Preliminary evidence from research and evaluation projects suggests positive changes in the pro-conservation attitudes and knowledge of participating students. To a lesser extent, this same research has also begun to indicate positive short-term changes in pro-conservation behaviours.

SUMMARY

Beginning teachers of science can feel they have a problem with identity and purpose. The problem is not so much in identifying what science *is* (e.g. see Bell et al., 2009; Bonney et al., 2009; Hodson, 1994), but rather identifying what a science education is *for* in the spaces and times in which it takes place (e.g. see Falk & Dierking, 2010; Henderson, 2012; Osborne & Dillon, 2008).

Throughout this chapter, we hope we have illustrated and provoked reflection on some key points:

- It is important to have a clear vision of what is possible and where that happens best.
- If a vision of what matters both *to* and *as* science education is not explicit, it is little wonder that we see issues of relevancy and authenticity regularly raised in science education (Braund & Reiss, 2006; Rennie et al., 2003; Stocklmayer, Rennie & Gilbert, 2010).
- Teachers need to find ways to make such a vision explicit in their practice or run the risk of falling back on traditional models or arrangements for science education (Dillon, 2016; Fallik, Rosenfeld & Eylon, 2013; Tunnicliffe, Lucas & Osborne, 1997).

In response to these issues, ensuring the curriculum emphasis in science education engages students for longer and in deeper and more authentic ways with particular learning objectives has become a common focus for innovation and curriculum renewal by governments, education departments and individual schools (DeWitt & Osborne, 2007; Fensham, 1985; Roberts, 1988). A near decade-old study by the National Research Council on learning science in informal environments (Bell et al., 2009) recognises that informal as well as free-choice learning activities (such as everyday experiences of walking in the park) contribute to people's knowledge of and interest in science. Less haphazardly, adults may satisfy their intellectual curiosity as well as fulfil a need for relaxation and enjoyment by visiting settings such as zoos and national parks, science centres and botanical gardens. Equally, studies from around the world show adults (parents, teachers, guardians, community leaders) may take children to these settings because they feel such experiences are worthwhile and fun, as well as educational, and that they and their children learn about science in the process (Kisiel et al., 2012; Mallapur, Waran & Sinha, 2008; Marino et al., 2010; Randler, Kummer & Wilhelm, 2012).

Nevertheless, while the focus of discussion on innovative science education approaches is usually centred on classroom-based, online or school environs (e.g. school gardens and grounds), a shift in consideration to diverse 'scapes' for science education opens up a range of opportunities for consideration (Dierking et al., 2002; Falk et al., 2007; Rickinson et al., 2004). In this chapter, we have suggested that there is significant opportunity in developing science education beyond the school gates, and that doing so—particularly by working with partner organisations supportive of the science education endeavour (Clayton, Fraser & Saunders, 2009; Falk, 2014; Patrick et al. 2007)—can help engender a sense of purpose that some science educators feel they sorely need.

Take-aways

Our particular focus has been to mobilise insights from considerations and reviews of learning at zoos (e.g. Dierking et al., 2004; Ogden & Heimlich, 2009; Scott & Matthews, 2011). However, in terms of socio-scientific issues, as examples of science institutions that often frame their mission as at the forefront of complex and highly relevant problems that science can help understand, some organisations—such as Zoos Victoria—design educational programs that try to solve them, for example via offering the Education for Conservation program. Studies demonstrate that the socio-scientific issue of wildlife conservation—a complex and growing issue across the world—drives the mission of more than 92 per cent of zoos (Patrick et al., 2007). Collaborating with a zoo can offer science teachers an opportunity for themselves and their students to engage with relevant real-world problems and, guided by and interacting with experts in conservation and education, to contribute meaningfully to solving them (Heimlich, 2010).

In the case of Zoos Victoria, but also more broadly, we have to recognise that each science-scape provider is located within a specific community context, and therefore the very issues that they work to solve are also often framed by that context. Schools choosing to collaborate with educators in a science-scape are able to explore real-world, localised issues that are likely to cross over with students' own personal experiences and interests. As a final example, the Seal the Loop conservation campaign developed by Zoos Victoria was in response to concern for the local Australian fur seal and other resident marine wildlife that were continually facing the threat of marine entanglement in Melbourne's Port Phillip Bay. Schools that have successfully collaborated with Zoos Victoria to protect these species have largely been located on Port Phillip Bay or near a Victorian local seal population, and have become personally invested in resolving this issue (see Case Study 10.1), providing an authentic and meaningful context and purpose to science education as a result.

CODA

As well as being a prophet of the digital age, and developing ideas such as the 'global village', Marshall McLuhan (1964) was also interested in the idea of 'feedforward'. Feedforward focuses on 'anticipating the effect of one's activities by acting as our own critic' (Coupland, 2010)—in other words, we might use ideas and provocations such as the notion of science-scapes to inform what we think and do about where science education happens, now and into the future.

DISCUSSION QUESTIONS

10.1 Carr and Cohen (2011, p. 175) argue that 'zoos are, today, identified as sites of conservation, research, education, and entertainment'. In your experience, and in developing your practice, which should get priority? Do you foresee any challenges—for example, how is an authentic and rich science education ensured?

10.2 Educators working for zoos may originally have started out as classroom secondary science teachers, or primary generalist teachers. Others come from a range of subject backgrounds or career areas, such as communication and media. They are typically expected to offer expert pedagogies for working with diverse groups of student learners who are being taught outside their familiar classroom setting. What would you identify as the key factors in preparing for, and developing, a career as a science educator working in a zoo?

10.3 Smith and Broad (2008) compare what can be learnt from the media (e.g. natural history documentaries) with incursions and excursions with zoo educators, using Weribee Zoo (part of Zoos Victoria) as a case study. A decade on, and in light of this chapter, is it, for example, an either/or scenario as they suggest, or is there something synergistic still to consider? How might the notion of a science-scape help to re-emphasise the value of informal settings for authentic science education?

10.4 A study by Randler and colleagues (2007) compares the cognitive and affective outcomes of adolescents learning at workstations at the zoo. Using the notion of science-scapes, revisit Table 10.1 and identify a variety of sites for the cognitive and affective outcomes of science education in your area of specialism or level of teaching. Now consider how you could *avoid* science learning and teaching being compartmentalised—in other words, what can be used to bridge the different domains and dimensions: the lifeworld of the student? What experiences would integrate the theory and practice of science learning? Focus on what is 'performed' at these sites (Penn, 2009).

10.5 Zoos can be emotionally charged, emotive and emotional places (Myers, Saunders & Birjulin, 2004). When science educators argue for the 'prioritisation of the affective' (Hodson & Reid, 1988), and for learners' direct participation in and experience of science, what words of warning or encouragement would you provide in relation to zoo-based education?

FURTHER READING

Association of Science and Technology Centers (ASTC) (2018). Website. Retrieved from www.astc.org.

Association of Zoos and Aquariums (AZA) (2018). Website. Retrieved from www.aza.org.

Bentleigh West Primary School Orangutan Celebration Day. Video. Retrieved from vimeo.com/172063323.

Community Conservation Campaign (2018). Teacher resources. Retrieved from www.zoo.org.au/education/teacher-resources.

Corrigan, D., Buntting, C., Jones, A. & Loughran, J. (2018). *Navigating the Landscape of Formal and Informal Science Learning Opportunities*. Dordrecht: Springer.

Eastern Barred Bandicoots Released at Woodleigh School. YouTube video. Retrieved from www.youtube.com/watch?v=lHgvl2IKt_Q.

Fighting Extinction School Community. Retrieved from www.zoo.org.au/education/fighting-extinction-schools.

Informal Science—www.informalscience.org—a repository of evaluation and research reports, many focusing on learning and education.

Zoos Victoria (2018a). The Global Environmental Education Partnership Education for Conservation Case Study. Retrieved from www.thegeep.org.

—— (2018b). Zoos Victoria Education website. Retrieved from www.zoo.org.au/education.

REFERENCES

Adams, J.D., Gupta, P. & DeFelice, A. (2012). 'Schools and informal science settings: collaborate, co-exist, or assimilate?' *Cultural Studies of Science Education*, 7(2), 409–16.

Amir, S. & Abraham, P. (2010). 'The role of animal-based attractions in ecological sustainability: Current issues and controversies'. *Worldwide Hospitality and Tourism Themes*, 2(3), 281–98.

Andersen, L.L. (2003). 'Zoo education: From formal school programmes to exhibit design and interpretation'. *International Zoo Yearbook*, 38(1), 75–81.

Association of Zoos and Aquariums (AZA) (2013). *The Accreditation Standards and Related Policies*. Washington, DC: AZA. Retrieved from www.aza.org/accreditation.

Ballantyne, R., Packer, J., Hughes, K. & Dierking, L. (2007). 'Conservation learning in wildlife tourism settings: Lessons from research in zoos and aquariums'. *Environmental Education Research*, 13(3), 367–83.

Ballouard, J.-M., Provost, G., Barré, D. & Bonnet, X. (2012). 'Influence of a field trip on the attitude of schoolchildren toward unpopular organisms: An experience with snakes'. *Journal of Herpetology*, 46(3), 423–8.

Bell, P., Lewenstein, B., Shouse, A.W. & Feder, M.A. (2009). *Learning Science in Informal Environments: People, Places, and Pursuits*. Washington, DC: National Research Council.

Bonney, R., Ballard, H., Jordan, R., McCallie, E., Phillips, T., Shirk, J. & Wilderman, C.C. (2009). *Public Participation in Scientific Research: Defining the Field and Assessing its Potential for Informal Science Education. A CAISE Inquiry Group Report*. Washington, DC: Center for Advancement of Informal Science Education (CAISE).

Braund, M. & Reiss, M. (2006). 'Towards a more authentic science curriculum: The contribution of out-of-school learning'. *International Journal of Science Education*, 28(12), 1373–88.

Bruni, C.M., Fraser, J. & Schultz, P.W. (2008). 'The value of zoo experiences for connecting people with nature'. *Visitor Studies*, 11(2), 139–50.

Bueddefeld, J.N.H. & Van Winkle, C.M. (2018). 'The role of post-visit action resources in facilitating meaningful free-choice learning after a zoo visit'. *Environmental Education Research*, 24(1), 97–110.

Carr, N. & Cohen, S. (2011). 'The public face of zoos: Images of entertainment, education and conservation'. *Anthrozoös*, 24(2), 175–89.

Clayton, S., Fraser, J. & Saunders, C.D. (2009). 'Zoo experiences: Conversations, connections, and concern for animals'. *Zoo Biology*, 28(5), 377–97.

Corrigan, D., Dillon, J. & Gunstone, R. (eds) (2007). *The Re-emergence of Values in Science Education*. Rotterdam: Sense.

Coupland, D. (2010). *Marshall McLuhan: You Know Nothing of My Work!* New York: W.W. Norton.

DeWitt, J. & Osborne, J. (2007). 'Supporting teachers on science-focused school trips: Towards an integrated framework of theory and practice'. *International Journal of Science Education*, 29(6), 685–710.

Dierking, L.D., Adelman, L.M., Ogden, J., Lehnhardt, K., Miller, L. & Mellen, J.D. (2004). 'Using a behavior change model to document the impact of visits to Disney's Animal Kingdom: A study investigating intended conservation action'. *Curator: The Museum Journal*, 47(3), 322–43.

Dierking, L.D., Burtnyk, M.S., Büchner, K.S. & Falk, J.H. (2002). *Visitor Learning in Zoos and Aquariums: A Literature Review*. Annapolis, MD: Institute for Learning Innovation.

Dillon, J. (2016). *Towards a Convergence Between Science and Environmental Education*. London: Routledge.

Dohn, N.B. (2013). 'Upper secondary students' situational interest: A case study of the role of a zoo visit in a biology class'. *International Journal of Science Education*, 35(16), 2732–51.

Dove, T. & Byrne, J. (2014). 'Do zoo visitors need zoology knowledge to understand conservation messages? An exploration of the public understanding of animal biology and of the conservation of biodiversity in a zoo setting'. *International Journal of Science Education*, Part B, 4(4), 323–42.

European Association of Zoos and Aquaria (EAZA) (2008). *Education Standards*. Amsterdam: EAZA. Retrieved from www.eaza.net/about/Pages/Key%20 Documents.aspx.

Falk, J.H. (2014). 'Evidence for the educational value of zoos and aquariums'. *WAZA Magazine*, 15, 10–13.

Falk J.H. & Dierking, L.D. (2010). 'The 95% solution: School is not where most Americans learn most of their science'. *American Scientist*, 98, 486–93.

Falk J.H., Reinhard, E.M., Vernon, C.L., Bronnenkant, K., Deans, N.L. & Heimlich, J.E. (2007). *Why Zoos & Aquariums Matter: Assessing the Impact of a Visit*. Silver Spring, MD: Association of Zoos & Aquariums.

Fallik, O., Rosenfeld, S. & Eylon, B.-S. (2013). 'School and out-of-school science: A model for bridging the gap'. *Studies in Science Education*, 49(1), 69–91.

Fennell, D.A. (2013). 'Contesting the zoo as a setting for ecotourism, and the design of a first principle'. *Journal of Ecotourism*, 12(1), 1–14.

Fensham, P.J. (1985). 'Science for all: A reflective essay'. *Journal of Curriculum Studies*, 17(4), 415–35.

Fraser, J. & Sickler, J. (2008). *Why Zoos and Aquariums Matter: Handbook of Research Key Findings and Results from a National Audience Survey*. Silver Spring, MD: Association of Zoos & Aquariums.

Grazian, D. (2012). 'Where the wild things aren't: Exhibiting nature in American zoos'. *The Sociological Quarterly*, 53(4), 546–65.

Griffith, J. & Symington, D. (1997). 'Moving from task-oriented to learning-oriented strategies on school excursions to museums'. *Science Education*, 81(6), 763–79.

Heimlich, J.E. (2010). 'Environmental education evaluation: Reinterpreting education as a strategy for meeting mission'. *Evaluation and Program Planning*, 33(2), 180–5.

Heimlich, J.E., Adams, J.D. & Stern, M.J. (2017). 'Nonformal educational settings'. In A. Russ & M.E. Krasny (eds), *Urban Environmental Education Review*. Ithaca, NY: Cornell University Press.

Heimlich, J.E. & Falk, J.H. (2009). 'Free-choice learning and the environment'. In J.H. Falk, J.E. Heimlich & S. Foutz (eds), *Free-choice Learning and the Environment.* Los Altos, CA: AltaMira Press, pp. 11–21.

Henderson, S. (ed.) (2012). 'Citizen science: New pathways to public involvement in research'. Special issue of *Frontiers in Ecology and the Environment*, 10(6).

Hodson, D. (1994). 'Seeking directions for change: The personalisation and politicisation of science education'. *Curriculum Studies*, 2(1), 71–98.

Hodson, D. & Reid, D. (1988). 'Changing priorities in science education'. *School Science Review*, 70, 101–8, 159–65.

Hopkins, C. (1991). 'Zoo education in Victoria, Australia'. *Journal of Museum Education*, 16(2), 17–18.

Jenkins, E.W. (1994). 'Public understanding of science and science education for action'. *Journal of Curriculum Studies*, 26(6), 601–11.

Kisiel, J., Rowe, S., Vartabedian, M.A. & Kopczak, C. (2012). 'Evidence for family engagement in scientific reasoning at interactive animal exhibits'. *Science Education*, 96(6), 1047–70.

Levinson, R. (2006). 'Towards a theoretical framework for teaching controversial socio-scientific issues'. *International Journal of Science Education*, 28(10), 1201–24.

Liu, B. & Reid. A. (2017). 'Lessons from a zoo-based conservation organisation'. *Eingana*, 40(2), 25–30.

Lucas, A.M. (1991). '"Info-tainment" and informal sources for learning science'. *International Journal of Science Education*, 13(5), 495–504.

Mallapur, A., Waran, N. & Sinha, A. (2008). 'The captive audience: The educative influence of zoos on their visitors in India'. *International Zoo Yearbook*, 42(1), 214–24.

Marino, L., Lilienfeld, S.O., Malamud, R., Nobis, N. & Broglio, R. (2010). 'Do zoos and aquariums promote attitude change in visitors? A critical evaluation of the American Zoo and Aquarium Study'. *Society and Animals*, 18(2), 126–38.

Matiasek, J. & Luebke, J.F. (2014). 'Mission, messages, and measures: Engaging zoo educators in environmental education program evaluation'. *Studies in Educational Evaluation*, 41, 77–84.

Maynard, L. (2017). 'Media framing of zoos and aquaria: From conservation to animal rights'. *Environmental Communication*, doi:10.1080/17524032.2017.1310741.

McLuhan, M. (1964). *Understanding Media: The Extensions of Man.* New York: Penguin.

Myers, O.E. Jr, Saunders, C. & Birjulin, A. (2004). 'Emotional dimensions of watching zoo animals: an experience sampling study building on insights from psychology'. *Curator*, 47(3), 299–321.

Neves, J.P.C. & Monteiro, R.C.R. (2014). 'How full is your luggage? Background knowledge of zoo visitors regarding sharks'. *Environmental Education Research*, 20(3), 291–312.

Ogden, J. & Heimlich, J.E. (2009). 'Why focus on zoo and aquarium education?' *Zoo Biology*, 28(5), 357–60.

Osborne, J. & Dillon, J. (2008). *Science Education in Europe: Critical Reflections*. London: Nuffield Foundation.

Packer, J. & Ballantyne, R. (2004). 'Is educational leisure a contradiction in terms? Exploring the synergy of education and entertainment'. *Annals of Leisure Research*, 7(1), 54–71.

Patrick, P.G., Matthews, C.E., Ayers, D.F. & Tunnicliffe, S.D. (2007). 'Conservation and education: Prominent themes in zoo mission statements'. *The Journal of Environmental Education*, 38(3), 53–60.

Pearson, E.L., Dorrian, J. & Litchfield, C.A. (2013). 'Measuring zoo visitor learning and understanding about orangutans: Evaluation to enhance learning outcomes and to foster conservation action'. *Environmental Education Research*, 19(6), 823–43.

Pearson, E.L., Lowry, R., Dorrian, J. & Litchfield, C.A. (2014). 'Evaluating the conservation impact of an innovative zoo-based educational campaign: "Don't Palm Us Off" for orang-utan conservation'. *Zoo Biology*, 33(3), 184–96.

Penn, L. (2009). 'Zoo theater's influence on affect and cognition: A case study from the Central Park Zoo in New York'. *Zoo Biology*, 28(5), 412–28.

Powell, D.M. & Bullock, E.V.W. (2014). 'Evaluation of factors affecting emotional responses in zoo visitors and the impact of emotion on conservation mindedness. *Anthrozoös*, 27(3), 389–405.

Randler, C., Baumgärtner, S., Eisele, H. & Kienzle, W. (2007). 'Learning at workstations in the zoo: A controlled evaluation of cognitive and affective outcomes'. *Visitor Studies*, 10(2), 205–16.

Randler, C., Kummer, B. & Wilhelm, C. (2012). 'Adolescent learning in the zoo: Embedding a non-formal learning environment to teach formal aspects of vertebrate biology'. *Journal of Science Education and Technology*, 21(3), 384–91.

Rennie, L. (2014). 'Learning science outside of school'. In N. Lederman & S. Abell (eds), *Handbook of Research on Science Education Volume II*. New York: Routledge, pp. 120–44.

Rennie, L.J., Feher, E., Dierking, L.D. & Falk, J.H. (2003). 'Toward an agenda for advancing research on science learning in out-of-school settings'. *Journal of Research in Science Teaching*, 40(2), 112–20.

Rennie, L. & McClafferty, T. (1995). 'Using visits to interactive science and technology centers, museums, aquaria, and zoos to promote learning in science'. *Journal of Science Teacher Education*, 6(4), 175–85.

—— (1996). 'Science centres and science learning'. *Studies in Science Education*, 27, 53–98.

Rickinson, M., Dillon, J., Teamey, K., Morris, M., Choi, M.Y., Sanders, D. & Benefield, P. (2004). *A Review of Research on Outdoor Learning*. Shrewsbury: Field Studies Council.

Roberts, D.A. (1988). 'What counts as science education?' In P. Fensham (ed.), *Development and Dilemmas in Science Education* (pp. 27–54). Philadelphia, PA: Falmer Press.

Roe, K. & McConney, A. (2015). 'Do Zoo Visitors Come to Learn? An Internationally Comparative, Mixed-methods Study'. *Environmental Education Research*, 21(6), 865–84.

Roe, K., McConney, A. & Mansfield, C. (2014a). 'How do zoos "talk" to their general visitors? Do visitors "listen"? A mixed method investigation of the communication between modern zoos and their general visitors'. *Australian Journal of Environmental Education*, 30(2), 167–86.

—— (2014b). 'The role of zoos in modern society: A comparison of zoos' reported priorities and what visitors believe they should be'. *Anthrozoös*, 27(4), 529–41.

Schultz, J.G.W. & Joordens, S. (2014). 'The effect of visitor motivation on the success of environmental education at the Toronto Zoo'. *Environmental Education Research*, 20(6), 753–75.

Scott, C.M. & Matthews, C.E. (2011). 'The "science" behind a successful field trip to the zoo'. *Science Activities Classroom Projects and Curriculum Ideas*, 48(1), 29–38.

Smith, L. & Broad, S. (2008). 'Comparing zoos and the media as conservation educators'. *Visitor Studies*, 11(1), 16–25.

Stocklmayer, S.M., Rennie, L.J., & Gilbert, J.K. (2010). 'The roles of the formal and informal sectors in the provision of effective science education'. *Studies in Science Education*, 46(1), 1–44.

Tunnicliffe, S.D., Lucas, A.M. & Osborne, J. (1997). 'School visits to zoos and museums: A missed educational opportunity?' *International Journal of Science Education*, 19(9), 1039–56.

Vernon, C. & Boyle, P. (2008). 'Understanding the impact of a zoo or aquarium visit'. *Connect*, April, 7–9.

Wagoner, B. & Jensen, E. (2010). 'Science learning at the zoo: Evaluating children's developing understanding of animals and their habitats'. *Psychology & Society*, 3(1), 65–76.

Westley, F. & Vredenburg, H. (1996). 'Prison or ark? The drama of managing the modern zoo'. *Studies in Cultures, Organizations and Societies*, 2(1), 17–30.

Wünschmann, S., Wüst-Ackermann, P., Randler, C., Vollmer, C. & Itzek-

Greulich, H. (2017). 'Learning achievement and motivation in an out-of-school setting: Visiting amphibians and reptiles in a zoo is more effective than a lesson at school'. *Research in Science Education*, 47(3), 497–518.

Yocco, V.S., Bruskotter, J., Wilson, R. & Heimlich, J.E. (2015). 'Why should I care? Exploring the use of environmental concern as a frame of communication in zoos'. *The Journal of Environmental Education*, 46(1), 56–71.

Chapter 11
Science and STEM education

DEBORAH CORRIGAN *and* **GREG LANCASTER**

LEARNING OUTCOMES
After reading this chapter, you should be able to:

- articulate the importance of science and STEM education in preparing students for the future workforce
- highlight the role of twenty-first century learning skills and what they mean for science learning and teaching
- consider the importance of curriculum integration and how this might play out in primary and secondary school contexts.

Science, technology, engineering and mathematics (STEM) are considered as key areas for ensuring economic prosperity into the future, both as individual disciplines and as a collective (Office of the Chief Scientist, 2015). STEM, as defined by Australia's Chief Scientist in 2016, includes:

- *Science*—the natural and physical sciences, agriculture, environmental and related studies. It should be noted here that health (and hence medicine) has not been included in this definition.
- *Technology*—defined as information technologies. In this instance technology is seen as providing tools and services to satisfy real world needs, and operates at the cross-section of science and society. Information and communication technologies is playing an ever-increasing role in our society and provides enabling capacity to other STEM disciplines.
- *Engineering*—defined as engineering and related technologies. Engineering draws on scientific, mathematical and technological knowledge and methods to design and implement physical information-based products, systems and services that address human needs, safely and reliably. Engineering takes into account economic, environmental, and aesthetic factors.
- *Mathematics*—seeks to understand the world by performing symbolic reasoning and computation on abstract structures and patterns in nature. It unearths relationships among these structures, and captures certain features of the world through the processes of modeling, formal reasoning and computation. (Office of the Chief Scientist, 2016, p. 2)

In this same report (Office of the Chief Scientist, 2016, p. 2), STEM education is defined as 'not merely impart[ing] content in these fields—it seeks to provide a framework in which new problems can be tackled'. STEM, in particular, has a critical role in relation to innovation and entrepreneurship. Underpinning and stimulating both innovation and entrepreneurship is creativity. This raises the question of where creativity is actually developed and nurtured within society generally, but especially in economics and education.

So what, if any, is the role of science and STEM education in this emerging directive? In our view, STEM education is pivotal to ensuring that our future thinkers are responsive and adaptable to emerging opportunities and challenges. Science is also important for the role it plays in developing expert thinking. This chapter will explore how the education sector can address the goals for the twenty-first century and what this will mean on the ground in primary and secondary school settings. There will be a particular focus on the development of twenty-first century skills—namely, critical thinking, creativity, collaboration and communication—through STEM education, even though STEM itself is not easily located within the school curriculum. Twenty-first century skills were developed by P21 in the United States in 2002—a coalition between the business community, education leaders and policy-makers (P21, 2002). Teachers and students can develop appropriate capabilities for their futures using disciplinary expertise such as science together with STEM in interdisciplinary spaces. Examples from professional practice and learning will be used to highlight the opportunities that can exist when the focus is on addressing the needs of learners, including the relevance of such learning.

THE ECONOMIC ENVIRONMENT: WHERE ARE WE AT, WHERE ARE WE HEADED?

In today's competitive global economy, STEM is considered a fundamentally important base for providing a skilled workforce to ensure future economic prosperity (Australian Academy of Science, 2011). According to Levy and Murnane (2007), the following global changes impact skills demand:

- *Automation*, where any routine work and thinking can be programmed to be completed by a computer. This includes information processing and simple pattern recognition as well as manufacturing tasks.
- *Globalisation*, which is the breaking down of economic, social and intellectual borders between nations. This has meant it is just as easy to collaborate across four continents as across four divisions within one organisation. Work practices such as offshoring, outsourcing and supply-chaining have meant a shift from vertical production (typical within a locality or at least a country) to horizontal collaboration (across the globe).
- *Workplace change*, which has meant less hierarchy and supervision, more autonomy and responsibility, and more collaboration. In addition, there is less predictability and stability as work is not organised around a person's specialty, but rather the problem to solve or task to achieve, and the end goal that needs to be realised (O'Toole & Lawler, 2006). The internet is already being used in a 'crowdsourcing' manner to find solutions to problems for a fee.
- *Personal risk and responsibility*, which have increased due to less job security (O'Toole & Lawler, 2006), increased personal responsibility for financial planning in retirement and ongoing health care.

In response to automation, Levy and Murnane (2007) propose that future workers need to build capacity in non-routine tasks—those tasks that cannot be broken down into a set of procedural rules and simple pattern recognition. They highlight two types of non-routine tasks: (1) expert thinking—solving unexpected problems that have no predictable or programmable rule-based solutions; and (2) complex communication—interacting with others to acquire information, explain it and/or persuade others of its importance.

Foundation skills such as reading, writing and mathematics provide a solid base for these tasks.

However, when we look at what such economies must look like, it is not obvious to many why STEM is a necessary foundation. For example, Australia's chief economist stated in the Australian Innovation System Report (Office of the Chief Scientist, 2015, p. 1) that:

> To continue to prosper, Australia must find new sources of wealth. As an advanced economy, we expect that further advances in national competitiveness and economic growth, including employment growth, will come primarily through innovation. Innovation is the core driver of business competitiveness and productivity. It supports economic growth, exports and job creation. Facilitating innovation involves enabling disruptive technologies and globalisation to access more opportunities for new products, new industries and new markets.

From an economic viewpoint, innovation appears to be the core driver for moving forward. It would therefore seem important that our students are well equipped to engage in the innovation processes they may encounter in the future. Interestingly, the term 'innovation' is used frequently, yet is rarely defined. Amabile (1988, p. 126) defines innovation as the 'successful implementation of creative ideas, particularly in an organization'. Importantly, innovation is linked to creativity, which Amabile (1988, p. 126) defines as 'the production of novel and useful ideas by an individual or small group of individuals working together', through the successful use or implementation of these ideas. Creativity has also been linked to twenty-first century learning skills along with critical thinking and problem-solving, collaboration and communication, which are often referred to in combination as the 4Cs. Each of the 4Cs is considered in more depth below.

CREATIVITY

As stated previously, creativity is about the production of novel and useful ideas (Amabile, 1988). It cannot be characterised as a lucky lightning bolt out of the blue. It requires both deep and technical expertise in one area and very broad knowledge in many apparently unrelated areas (Levy & Murnane, 2007). The more ideas you have, the more likely you are to have a truly creative insight, so success in creativity is linked to the quantity of output. Additionally, creativity is closely linked with conscious effort, hard work and persistence. Creativity depends on being able to combine quite disparate elements in new ways that are appropriate for the task or challenge at hand. It relies heavily on synthesis and the ability to see patterns where others may see chaos. Creativity will only happen in circumstances where the creator is allowed to fail many times in order to succeed—even if only once. It is somewhat ironic that the concept of failure is often not supported in many modern education systems, yet is so fundamental to the highly desired capability of creativity. People tend to feel most creative when they are intrinsically (internally/personally) motivated and satisfied by the challenge of the work itself. Extrinsic (external/outside) motivation is rarely conducive to creativity (Amabile, 1988).

At this point, it makes sense to take some time to sit and reflect on your own education and think about the opportunities you have had for creative expression in your own experience. As teachers, it is equally important to think about the opportunities we will provide for our students to engage in truly creative experiences. For example, creative opportunities in science usually exist in a team or collaborative environment, whereas creative events in the arts tend to be more individual in nature. So creating opportunities for students to work in teams, to brainstorm together, to explore all possibilities in an environment where learning from failure is an essential element for learning is one consideration for teachers.

CRITICAL THINKING

Critical thinking requires three predominant types of thinking: (1) reasoning, (2) making judgements and decisions, and (3) problem-solving (Willingham, 2007).

In this instance, reasoning requires the analysis and synthesis of evidence and then some evaluation of that evidence. This subsequently results in the drawing of conclusions based on the evidence that has been gathered. As highlighted in Chapter 6, evidence is data that have been given some value—it is seen as having worth—as opposed to data, which may be discarded. Similarly, data of different types can be attributed more value than other types of data. This is often the case when considering statistical data compared with anecdotal data.

The evaluation of evidence relies on making judgements and decisions, as it is possible that two scientists given the same data will reach differing conclusions due to their own experiences and contexts. Additionally, each of these scientists would be expected to acknowledge the other's alternative explanations and viewpoints, so making some judgements and decisions about these differing opinions is necessary.

Problem-solving in relation to critical thinking pertains to non-rule based approaches to solving issues, which is often not the type of problem-solving experienced at school. For example, problem-solving in mathematics is often of the type seen in Figure 11.1.

In this problem, it is important to understand the BODMAS (which stands for Brackets, Orders of, Division, Multiplication, Addition, Subtraction) rule, which dictates the order in which this problem

Given:

A = B + 3 (4 – C)

If B equals 5 and C equals 2, what is the value of A?

a) 7

b) 11

c) 12

d) 17

Figure 11.1 Rule-based problem-solving

needs to be solved. So brackets first means 4 – C, which is 4 – 2 = 2. The multiplication is next, so 3 x 2 = 6; then additions, so B + 6, which is 5 + 6 = 11. The solution is therefore b) 11.

If you do not know, or have forgotten, this rule (as many of us would!), then it is not possible to solve this problem. But this type of problem can easily be solved by a computer as you can develop a code to follow these 'rules'.

The type of problem-solving that forms part of critical thinking relates to those problems that cannot be solved by rules. This is because they often have multiple solutions and judgements that need to be made based on context, regarding which solution(s) make sense. An example is provided in Figure 11.2. Drawing upon a standard rule-based chemistry problem (where you would be asked to calculate the volume of $CO_{2(g)}$ at STP (standard temperature and pressure)), it has been changed into an open-ended problem with multiple solutions. This requires students to connect many ideas they have in order to generate possible answers (some of which are indicated in the figure).

Of course, problem-solving does not have to be numerical in nature, which is more often the case in science and STEM contexts.

An important aspect of critical thinking is the development of *expert thinking* (Levy & Murnane, 2007). In Figures 11.1 and 11.2, there is some need for expert knowledge in mathematics or

Example

A 10 g lump of marble is dropped into a beaker containing 250cm^3 of 2M HCl. Given the information below, find out as many things as you can about this situation.

$$CaCO_{3\,(s)} + 2HCl_{(aq)} \rightarrow CaCl_{2\,(aq)} + H_2O_{(l)} + CO_{2\,(g)}\uparrow \quad \Delta H = -101 \text{ kJ/mol}$$

A_m values: Ca = 40.1, C = 12, O = 16, H = 1, Cl = 35.5

Some possibilities you may have come up with are:

- mass of all products
- volume water produced
- volume of carbon dioxide at STP
- find concentrations of hydronium ions, calcium chloride, chloride ions
- find pH
- number of moles of calcium carbonate in excess
- mass HCl consumed
- heat produced
- temperature rise in beaker (may need to give 4.18 J/mol, ml).

Figure 11.2 Converting a rule-based problem to a non-rule based problem

chemistry. Experts' ability to think and solve problems depends strongly on a rich body of knowledge that is highly organised and connected into a schema, and where concepts are linked to the underlying 'big ideas'. For example, one of us (Corrigan) is a chemist and can group school chemistry knowledge around the three big ideas of (1) the notion of particles, (2) that these particles can be rearranged and (3) when they rearrange and pack together they consume or emit energy. All other chemistry-based concepts can be linked to these three big ideas, which fosters a deep understanding in relation to these three ideas. An important aspect of expert thinking is pattern recognition. Experts work from a highly organised knowledge base (a *schema*), to which they can easily connect new knowledge. They do this by looking for patterns as ways to connect new information (e.g. what is similar or different). Experts are also *metacognitive*—that is, they think about the way they think. This is not only important for their personal expertise, but also when dealing with experts from different disciplines, as they can begin to examine different ways of thinking and this assists in appreciating alternative explanations and viewpoints. Table 11.1 begins to detail the different ways experts in science and technology-engineering think. In considering this table, there is an opportunity for you to extend this to the ways experts in the fields of mathematics and information and communication technology (ICT) might think and act as well. The benefit for teachers of engaging in this thinking about these areas is that it enables them to articulate their own understandings about each of these disciplines and to look for similarities and differences in the way they need to think in each of them. Furthermore, teachers can work with their students to generate a class understanding of the thinking required in each of these disciplines, which allows students to begin to understand that there are similarities and differences in the thinking, even when similar words like 'reasoning' are used.

COLLABORATION

Collaboration is also an important capability to bring to the twenty-first century. It involves the ability to relate well to others. This might require an individual to be empathetic or take the role of the other person in imagining a situation from their perspective. In this circumstance, it is also important to effectively manage your own emotions.

Collaboration, as it is perhaps most commonly understood, also requires the ability to cooperate. This may mean you can not only present your own ideas, but you are willing to listen to the ideas of others too. To be truly collaborative, you need to be able to understand the dynamics of debate and follow an agenda, construct

Table 11.1 STEM as a way of thinking and acting

Science	Technology-engineering	Mathematics	ICT
Takes nature apart in order to understand or explain it	Puts nature together in order to create something novel		
Interested in natural phenomena	Interested in artificial things		
Essentially analytical in its thinking	Interested in essentially synthetic problems		
Interested in being able to generalise knowledge by inventing concepts and laws and even ideal situations like ideal gases and frictionless or unbending surfaces	Interested in specific knowledge—knowledge that has a bearing on a real, specific context and that provides detail about a specific problem		
Often driven by knowing for its own sake, by a fascination with natural phenomena	Always has a human need or opportunity in mind		
Basically comfortable with notions like 'discovering' or 'uncovering' nature	Basically comfortable with notions of 'design' and 'invention'		

tactical or sustainable alliances, be able to negotiate and make decisions that allow for differences in opinion.

Finally, collaboration requires the ability to manage and resolve conflicts. This will require an ability to analyse issues and interests at stake (e.g. power, recognition of merit, division of work, equity), the origins of the conflict, the reasoning of all sides, and that there are different possible positions. Other abilities, such as identifying areas of agreement and disagreement, reframing the problem, prioritising needs and goals, and making decisions about what you are willing to give up and under what circumstances are also necessary.

In many school situations, group work is often used as a synonym for collaboration. It is important to note that this is in fact rarely the case, as group work is often quite superficial, with little or no purpose to the grouping. It does little to develop the complex range of skills that are required to truly collaborate. Often groupings in schools are random or based on competency in a particular skill rather than developing relational skills, conflict-negotiation skills and acceptance of a range of possible views.

COMMUNICATION

The development of complex communication skills is an essential capability, both now and into the future. As highlighted in Chapter 6, scientists can no longer just rely on communicating with each other. They have a range of audiences with which they need to communicate and there are many different genres they will need to use. Hence, having capability in a range of communications (such as oral, verbal and visual) is important; however, equally important are the non-verbal and tacit (hidden) communications between people collaborating on problems and issues. The requirements around what it means to collaborate also means that collaborators need to be *in tune* with each other. This is not the same as *group think*, but rather involves being aware of what the people you are collaborating with are saying and thinking, and valuing this diversity.

WHAT DOES THIS MEAN FOR SCHOOL EDUCATION?

Currently in primary school settings, literacy and numeracy typically are the focus of a significant amount of attention during the school week. All other aspects of the curriculum are often dealt with in an integrated manner—in the sense that there is a bringing together of previously separate (or siloed) learning areas. In terms of the types of thinking required from STEM education, the notion of an integrated curriculum provides enormous opportunities for working in the spaces that exist between the disciplines or key learning areas.

While there are numerous recorded instances of students in primary schools working together in integrated ways, there are fewer instances of this happening in secondary school settings. However, when it does occur, it is more likely that this approach to curriculum will be evident in the middle school years (often Years 5 to 8), where there are fewer structures in place forcing a uni-disciplinary approach than in the post-compulsory years (Years 11 and 12) (Rennie, Venville & Wallace, 2018). Nevertheless, many researchers have tried to categorise these integrated approaches in a variety of ways, which Meeth (1978) summarises as follows:

- *A cross-disciplinary approach* is where one discipline is examined from the perspective of another (e.g. science of music).
- *Multidisciplinary approaches* have two or more disciplines contributing to the study of a single issue or problem, while each discipline retains its own identity (e.g. physics and mathematics concepts being used to design a solar cooker).
- *Interdisciplinary approaches* blur the boundaries between the disciplines.
- *Transdisciplinary approaches* involve disciplinary boundaries being dissolved. While this remains the most difficult approach, it is the preferred option to solve major global issues, such as the recent Ebola threat.

In relation to science and STEM, Rennie, Venville and Wallace (2012a) categorise ways in which schools approach integrated curriculum. This categorisation is based on a description of its approach and purpose. They describe six kinds of approaches to integration: (1) synchronised, (2) thematic, (3) project-based,

Table 11.2 Summary of approaches to curriculum integration

Approach to integration	Description of integrated approach with example
Synchronised	Teachers identify skills, knowledge or understanding that are common to particular topics in two or more subjects and teach these topics separately but, in parallel, explicitly drawing cross-curricular links and reinforcing concepts. Example: For a Year 9 applied class, the science and geography teachers identified areas of curriculum overlap and collaborated to teach a five-week energy topic in parallel, making clear the common concepts and using common assessment items.
Thematic	Teachers work collaboratively (perhaps in learning teams) to organise the curriculum around a local or global topic (e.g. insect infestation, the Olympic Games). They teach their subjects separately in a complementary way, making connections back to the theme. They may come together in a culminating event incorporating aspects of all learning areas. Example: All five Year 8 teachers worked as a team to run integrated projects (e.g. environmental theme day), incorporating aspects of all learning areas.
Project based	The focus here is on a designated task requiring knowledge and skills from more than one subject area for its completion. Projects are often technology or engineering based, requiring construction of some kind. Example: A Year 9 technology class was set a bridge-building project, which incorporated knowledge of science, mathematics, engineering, design and construction. The aesthetics of the bridge were judged by the English teacher.
Cross-curricular	These approaches are usually based around overarching skills or competencies, such as literacy, numeracy, ICT skills, or perhaps social skills such as cooperation or environmental responsibility. Integration occurs when these skills are the focus of more than one subject at the same time.

Approach to integration	Description of integrated approach with example
	Example: A school-wide literary focus aimed at assisting Aboriginal students to learn English, which underpinned all subject areas and was supported by cross-curricular initiatives, such as horticulture relating to the school garden.
School specialised	When a school has a long-term focus on a specific area, such as aviation studies, teachers in the other subject areas may tailor their courses to have explicit links to this specialisation. Example: A coastal high school developed a marine studies specialisation. Teachers in each of the 'core' subject areas taught one specified unit of marine studies in each of Years 8–10.
Community focused	Here, a significant community issue, such as maintenance of a local wetland, becomes the curriculum focus, and teachers orientate their subject teaching to help students understand the issue from different perspectives and seek potential ways of dealing with it. Example: Two Year 8 classes focused a term's work in science and social studies on access for the disabled in community venues and activities.

Source: Based on Rennie, Venville & Wallace (2012a, Table 2.1).

(4) cross-curriculum, (5) school-specialised and (6) community-focused programs. These approaches are outlined in Table 11.2.

Rennie and colleagues (2012a) found that synchronised and thematic approaches were usually multidisciplinary, as teaching within subjects prevailed. Project-based approaches were more interdisciplinary, as the subject boundaries were blurred, which was also the case in cross-curricular approaches. Subject-specialised approaches were at least multidisciplinary, but could be inter-disciplinary, while community-focused approaches were often interdisciplinary but had the potential to be transdisciplinary because meaning-making was focused on an issue, not on the content of each subject. Following on from this work, Rennie, Venville and Wallace (2012b) also identified that a small and stable learning environment, effective leadership, teacher-team activities linked to the classroom, time for in-school planning, a flexible timetable and the ability to make community links were paramount attributes for the success of integrated approaches.

It is interesting to ponder what this all means in terms of our practice as primary or secondary teachers of science. In thinking in these ways about science and STEM learning and teaching matters, Rennie, Venville and Wallace (2018, pp. 91–2) point out:

> We live in a connected, global community, and students should experience a school education that reflects this reality. In other words, the curriculum, particularly curriculum that is related to science, technology, engineering, and mathematics (the STEM subjects), must provide opportunities for students to explore and experience the kind of connectedness that reflects life outside of school. Most often, however, we find that schooling in these subjects does not encourage the embedding of the curriculum in real-world, authentic contexts, and this is especially so in secondary schools. Instead, current educational policy favours disciplinary approaches to knowledge that are translated into curriculum documents that espouse educational outcomes narrowly focused on what is readily measurable or amenable to achievement testing. Left out are the less cognitive- based outcomes that are so important in building the range of citizenship skills that we expect students to possess by the end of their school years.

DISCUSSION QUESTIONS

11.1 What impact do the constructs of twenty-first century learning skills and STEM education have on your thinking about science learning and teaching and, in particular, your role in this?

11.2 Choose one of the 4Cs and consider how you might enact this twenty-first century learning skill as part of your approach to science learning and teaching.

11.3 Which approach to curriculum integration most appeals to you? Why?

11.4 How do you now view yourself and your role in preparing students for the future as a teacher of science?

11.5 Which chapters of this book were you most drawn or connected to and why?

FURTHER READING

Felder, R.M. & Brent, R. (2016). *Teaching and Learning STEM: A Practical Guide.* San Francisco: Jossey-Bass.

Jorgensen, R. & Larkin, K. (eds), *STEM Education in the Junior Secondary: The State of Play.* Dordrecht: Springer.

REFERENCES

Amabile, T.M. (1988). 'A model of creativity and innovation in organisations'. *Research in Organisational Behaviour,* 10, 123–67.

Australian Academy of Science (2011). *Australian Science in a Changing World: Innovation Requires Global Engagement.* Position paper. Canberra: Australian Academy of Science. Retrieved from www.science.org.au/reports.

Levy, F. & Murnane, R.J. (2007). 'How computerized work and globalization shape human skill demands'. In M.M. Suarez-Orozco (ed.), *Learning in the Global Era: International Perspectives on Globalization and Education.* Berkeley, CA: University of California Press, pp. 158–76.

Meeth, R.L. (1978). 'Interdisciplinary studies: A matter of definition. *Change,* 10(7), 10.

Office of the Chief Scientist (2015). *The Importance of Advanced Physical and Mathematical Sciences to the Australian Economy.* Canberra: Commonwealth Government.

—— (2016). *Australia's STEM Workforce Report*. Canberra: Commonwealth Government.

O'Toole, J. & Lawler, E.E. III (2006). *The New American Workplace*. New York: Palgrave Macmillan.

Rennie, L.J., Venville, G. & Wallace, J. (2012a). *Knowledge that Counts in a Global Community: Exploring the Contribution of Integrated Curriculum*. London: Routledge.

—— (2012b). *Integrating Science, Technology, Engineering, and Mathematics: Issues, Reflections and Ways Forward*. New York: Routledge.

—— (2018). 'Making STEM curriculum useful, relevant and motivating for students'. In R. Jorgensen & K. Larkin (eds), *STEM Education in the Junior Secondary: The State of Play*. Dordrecht: Springer, pp. 91–109.

Willingham, D.T. (2007). 'Critical thinking: Why is it so hard to teach?' *American Educator*, 31(2), 8–19.

INDEX

4Cs 306–13

accountability 81–2
adaptive 68
affective 69
alternative conception 16, 29–31, 35–41, 42, 44–5
assessment
  aligning with quality learning and teaching 200–1
  as learning 183
  brainstorming 185–6
  clarifying expectations 198–200
  feedback 198–200
  learning logs 191
  overview 19, 177–83, 201–2
  portfolios 190
  practical work 193–5
  predict, observe, explain (POE) 192–3
  slowmation 195–6
  strategies 185–201
  tests and examinations 196–8
  think-pair-share 186
  types and purposes 183–5
  visual representations 187
    concept maps 187–9
    Venn diagrams 189–90
Association of Zoos and Aquariums (AZA) 280
attributes 32
attitudes 27, 69, 100, 103, 121, 124, 128, 180, 182, 290
Australian Curriculum 101–10
  content 101–4
  inclusivity, importance of 102–3
  science as a human endeavour (SHE) 106, 108
  science inquiry skills (SIS) 106–7, 108
  science understanding (SU) 106, 108
  structure 104–6
Australian Innovation System Report 306
Australian Institute for Teaching and School Leadership (AITSL)
  Professional Standards 249–53
Australian Professional Standards for Teachers 250–3

Australian School Science Education National Action Plan 2008–12 102
authentic 58
automation 305

banking model 215
beliefs, impact of 6–7
big ideas 59–60
Botanic Gardens 279

case studies 62–6, 72–5, 97–9, 114–17, 145, 147, 168–70, 207–9, 224–8, 263–5, 284–6, 287–8
causal relationship 153
Centre for Education and Research in Environmental Strategies (CERES) 279
claim 149
cognitive 69
collaboration 311–13
communication 313
community focused integration 316
competence 58
complex contexts 60–1
concept 193
conceptual change 39
constructivism 16, 31–5
contemporary versus traditional science 104
context
  complex 60–1
  features at play 61
  impact of 6–7
  role of 17, 57–8, 61–8, 82–3
  teaching considerations 68–72, 78–82
    accountability 70–1
    expectation and risk, managing 70
    school structures, reality of 69–70
continuity
  conceptual and contextual, difference between 70
continuous model 212
control group 153
correlational investigation 7
creativity 126, 307–11
critical moments 61
critical thinking 49, 308–11
cross-curricular integration 315
CSIRO Discovery Centre 279
curiosity 126
curriculum 17–18, 88
  aims 96
  Australian 101–6
  case study 97–9
  experienced/realised 90, 92–4, 108–10
  hidden 109
  implemented 90, 92–4, 106–8, 110
  inclusivity 102–3
  intended 89–90, 92–4, 101–6, 110
  meaning 90–2, 110
  null 109
  positive consequences 103
  possibilities
    sources/possibilities/pressures 99–100
    what could be included 100–1
  pressures 95
  purposes of 94–6
  relevance 103–4
  societal demands 95–6
  syllabus and 92
Curriculum and Standards Framework II (CSF II) 4

dependent variable 151

descriptive investigation 120
diagnostic assessment 184
different ways of knowing 31, 42–8
discernment 58
disposition 58

early years teaching 13
effective learning
  essential elements 48
empirical data 119
empiricism 127
engagement 66–8
  generating a need to know 67–8
  incorporating other skills and learning areas 68
  student-led exploration 66–7
epistemic 125
evaluative assessment 185
everyday interpretations 67
expectation, managing 70, 81
experimental science 120
extrinsic 76

fieldwork 120
fluency 58
formative assessment 184

globalisation 305
graduate teacher reflections 15

hidden curriculum 109
hypothesis 120

in-service teacher 6
inclusivity 102–3
independent learner 53
independent variable 151
independent verification 120
Indigenous perspective 29–30, 42–4, 45–8
individual concepts 59
inference 119
influences of society 78
informal learning settings 234
  beyond field trips 281–3
  case studies 263–5, 284–6, 287–8
  'feedforward' 293
  informal science institutions 279
  key considerations 267–70
  learning outcomes 288–90
  overview 20, 290–3
  program collaboration 263–7
  'scape', influence of 270–5
  science 'scapes' 275–80, 283
  zoos, collaboration with 283–8
  Zoos Victoria *see* Zoos Victoria
innovation 306–7
interactive dialogue 28
intrinsic 76
investigation 18–19, 88
  articulating expectation 155–6
  bias, reduction of 152
  case studies 145, 147, 168–70
  communicating with others 165–7
  defining a problem 151–2
  examining results 159–63
    controls and variables 160–2
    descriptive statistics 159–60
    tables and graphs 162–3
  investigating what is already known 154
  investigative approaches 143–9
  observing 154–5
  pedagogical approaches 167–74
    restructuring investigation 168–71
    thinking tasks 171–3

posing questions 152–4
processes and types 156–9
cross-sections 158
field sketching and photography 157
investigating opinions 159
sampling techniques 158
transects 158
water safety and quality 158
reflecting on findings 163–5
rules 144
scientific method 150–1
types 152–4
what constitutes 149–67
iterative process 146

knowledge, impact of 6–7

law 122
learner
paying attention to 28–9
learner of science vs teacher of science 11–12
learning
active process 53–4
changing thinking 48
conditions 50–1, 75–8
ownership of 52
learning outcomes 32
linear practice 146
linking knowledge 72–7
literature review 144

metacognitive 183
misconception 36
monitoring and assessment 8

NAPLAN 65, 71
nature of science (NoS) 18, 88
building on knowledge, skills and understandings 133–5
case study 114–17
communication 123–5
considering context 135–6
fostering inclusive learning environments 136–8
human endeavour 121–3
implications 129–40
meaning 113–14, 117–18, 128–9
science and school science
gap between 129–31
implications of reducing gap 132–8
negotiating the gap 131–2
values 125–7
what is 118–19
world perspective 119–21
need to know 77–8
null curriculum 109

open-mindedness 126
open reporting 120

paradigm 164
Parks Victoria 279
parsimony 126
pedagogical content knowledge (PCK)
case study 207–9, 224–8
content representation (CoRe) 216–29
making PCK useable 216–29
overview 19, 178, 205–9, 230–1
Professionals and Pedagogical-experience Repertoires (PaP-eRs) 216, 224–9
thinking about learning and teaching 210–13
conceptual change model 210–13

predict, observe, explain (POE) 211
vision for future professional learning 229–30
what is 214–16
pedagogy 240
peer review 120
peer-teaching task 4–5
personal risk and responsibility 305
pre-service teacher 4–5, 8–10
reflections 9
preconceptions 39
primary school teaching 13
professional knowledge 254
social construction of 254
professional practice
career stages 245–9
codes of conduct, adherence to 244
complexities 241–3
enhancement 253–8
evidence of proficiency 252–3
expertise, journey to 238–40
facets 238
growth and development 243–5
overview 19–20, 234–8, 258–9
professional standards 249–53
teacher knowledge of practice 240–1
project based learning 315, 316

questionnaires 159

random sampling 158
rational thinking 127
reflection 49
reflective scepticism 96
relevant science education 75–6, 78–80
definition 75–6
risk, managing 70, 81
role-play 4–5

scepticism 127
school specialised integration 316
school structures, reality of 69–70, 80–1
Science Continuum 134–5
science curriculum *see* curriculum
science investigations *see* investigations
science phenomena 31–2, 33–4
Science Teaching and Learning (STaL) program 63
Scienceworks 279
scientific beliefs 7
scientific interpretations 67
scientific method 7
secondary school teaching 13–14
Shape of the Australian Curriculum: Science 104
simple applications 60
social constructivism 33
STEM education 20–1, 234
collaboration 311–13
communication 313
creativity 307–8
critical thinking 308–11
definition 302–4
economic environment 304–7
engineering 303
4Cs 306–13
integrated approach 313–17
mathematics 303
role 304
science 303
technology 303
way of thinking and acting, as 312
student-centred perspective 81

summative assessment 184
surveys 159
syllabus 92
synchronised integration 315

taxonomy 155
teacher development *see* professional practice
teacher identity 10–11
teacher knowledge of practice 257
teacher of science vs science teacher 14–15
teachers as learners 49
thematic integration 315
theory 122
triangulation 154

Universal Design for Learning (UDL) 137–8

validity 149
variable 151
Victorian Space Science Education Centre (VSSEC) 279

'ways of knowing' framework 58–9, 61
what the learner brings 16–17
  alternative conception 16, 29–31, 35–41, 42, 44–5
  constructivism 16, 31–5
  different ways of knowing 31, 42–8
  teaching implications 48–53
workplace change 305

zoos
  informal learning spaces, as 283–8
Zoos Victoria 279
  Education for Conservation program 284–6
  Seal the Loop conservation campaign 263–5, 293

For Product Safety Concerns and Information please contact our EU representative GPSR@taylorandfrancis.com
Taylor & Francis Verlag GmbH, Kaufingerstraße 24, 80331 München, Germany

www.ingramcontent.com/pod-product-compliance
Lightning Source LLC
LaVergne TN
LVHW011112300926
849160LV00008B/356

* 9 7 8 1 7 6 0 2 9 6 8 8 9 *